The Physics

of

Empty Space

Dennis Morris

(January 2015)

© Dennis Morris

Published by: Abane & Right

31/32 Long Row

Port Mulgrave

Saltburn

TS13 5LF

United Kingdom

01947 840707

dennis355@btinternet.com

14th January 2015

Revised July 2015

Contents

Contents

Contents

Contents

What is in this Book

This book presents the nature of the 4-dimensional space-time in which we sit. We derive the existence of our 4-dimensional space-time with its properties including general relativity from the real numbers and the finite groups. We also derive classical electromagnetism. In the course of doing this, we discover prospective explanations to some other questions such as why there are three generations of particles, why we see no gravitons, and what might be quantum gravity.

One result shown this book is that, like the physical world, mathematics comes in both a quantum form and a classical form.

Quantum mathematics is no more than different types of number; that is different types of division algebras like quaternions, $\mathbb{H}$, or the Euclidean complex numbers, $\mathbb{C}$. We now know these to be spinor spaces.

Classical mathematics emerges from quantum mathematics by adding isomorphic division algebras to form an 'expectation algebra'; this is similar to the way we take expectation values in quantum theory. Classical maths is set over $\mathbb{R}^n$. This means that the two types of physics, quantum physics and classical physics, match the two types of mathematics.

Introduction

Science explains the world we observe. General relativity explains gravity. QFT explains the interaction of particles and the nature of atoms. The other sciences explain chemistry or biology or geology and much more of the world which we observe. However, there seems to be hole in our sciences in that no science explains the nature of the empty space-time which we clearly see around us. Surely the nature of empty space is a part of physics, but modern physics does not seriously address empty space. There are questions about empty space which are rarely asked. There are questions about empty space which, though asked, seem to have no answer other than a wild guess. There are questions about empty space that no-one realises are about empty space but are thought to be about something else. Some examples of such questions are:

1) Why do I have both a left hand and a right hand?

I cannot change my left hand into my right hand by any combination of rotations and translations. If I cannot change a left hand into a right hand by rotating and translating it, then perhaps the left hand is in a different space from the right hand. Surely, if two seemingly identical objects are in the same space, then one can be replaced by the other with no more than rotation and translation; that seems sensible[1]. In technical words, why do we have orientability in the space-time in which we sit? Orientability seems to separate objects within a space into two types – the left-handed type and the right-handed type.

2) Why, at a given time, can I know the spin of an electron in only one of the three spatial directions?

Within atomic physics, we observe that we can know only one component of angular momentum at a time. Any 5-year old will tell you that, in a single geometric space, this is not natural. We

---

[1] A weird form of sensible, perhaps.

have become so used to this idea that we lack the wisdom of the five year old to see that it is not natural. Humankind has constructed a mathematical formulation (quantum mechanics) based on non-commuting operators to describe what we observe. We associate three non-commutative operators with the three components of angular momentum, but we know that these three operators are the same single operator written in three different bases. Is the space-time in which we sit three spaces written in different bases and we observe in only one base at a time? Perhaps this is the reason for the mutual exclusion of observation of angular momentum components or perhaps there is another explanation.

3) Why is our space-time 4-dimensional?
4) Why is the distance function of our space a quadratic form with signature $(+,-,-,-)$?

The last of these questions might be asking, "Why are three of the dimensions of our space-time spatial and one temporal? Why is there an imbalance in the number of plus and minus signs in the distance function of our space-time?" but it is also asking why the distance function is not a cubic form or some wild mixture of expressions taken from an infinite number of possibilities. Question three is asking, "Why there are four independent variables in the distance function of the space-time in which we sit?"

5) How does empty space expand as astronomers observe it to be doing?

How can emptiness become more empty, and why does the rate of expansion appear to be changing?

6) Why can we not rotate 4-dimensionally in our space-time?

We have physical familiarity with only rotation in 2-dimensional planes. We are familiar with two types of 2-dimensional rotation; these are the Euclidean rotation of the flat plane and the velocity boost of space-time. These two types of rotation are expressed as

2×2 rotation matrices containing the appropriate trigonometric functions. Within mathematics, there are 3-dimensional rotations expressed as 3×3 rotation matrices and 4-dimensional rotations expressed as 4×4 rotation matrices[2]. One example of a 4-dimensional rotation is the 4×4 quaternion rotation matrix. Why do we, in this 4-dimensional space-time, not have a 4-dimensional rotation expressed as a 4×4 rotation matrix?

7) If empty space is 'empty', how is there more of it between the planet Pluto and the Earth than there is between the planet Mars and the Earth?
8) Why is quantum mechanics weird? Why is the physical universe separated into two parts, the quantum part and the classical part?

Quantum mechanical phenomena such as non-locality, instantaneous collapse of the spatially extensive wave function, superposition, and why the universe is not deterministic might be phenomena associated with the nature of empty space. It seems that the separation of physics into quantum physics and classical physics goes hand in hand with the separation of empty space into two types, upon which more will be presented later in this book.

There is no shortage of observations of empty space to explain. The distance function is well known and that space-time is 4-dimensional is without doubt, yet it goes unexplained. In this work, your author presents a mathematical understanding of empty space that provides prospective answers and insights into the above questions. The answers require an interpretation of the mathematics. Your author believes the mathematics is correct, but we can never be certain that any interpretation is correct; we can only assert its correctness and hope to convince others of that correctitude, or, perhaps, let ourselves be persuaded against that interpretation by others. Still, without interpretation, mathematics is no more than chalky squiggles on a

---

[2] See Dennis Morris. Complex Numbers The Higher Dimensional Forms ISBN: 978-1508877499

blackboard. Of course, the reader will choose whether or not the interpretation is convincing.

With every interpretation, there is a danger that the interpretation will be taken by future physicists as being reality. As it was once fashionable to think of time and space as Newton thought of time and space, so it might become fashionable to think of time and space as it is presented within this work; indeed, to make it fashionable is your author's intention, but we must be aware that such fashions are only fashions and can often be a block to the deepening of human understanding. Underlying any physical theory, there are unstated and often unseen assumptions that obscure understanding and befuddle progress. Often, it is the realisation that an 'absolute truth' is actually an assumption that leads to breakthroughs in human understanding. Perhaps your author has unwittingly made unseen assumptions in this work.

Chapter 1

Overview of the Nature of Space

Within mathematics and physics, there is a generally held, but little questioned, view that n-dimensional space is n 1-dimensional spaces fixed together somehow at right angles to each other. This view is very useful in applied mathematics where we take, say, both supply and demand to each be real numbers. This view is very useful for designing bridges or planning space-flights. We should not decry the usefulness of this $\mathbb{R}^n$ view of n-dimensional space; general relativity is formulated over $\mathbb{R}^4$ and economists and engineers would be lost without $\mathbb{R}^3 \, \& \, \mathbb{R}^2$, but there are questions about applying this $\mathbb{R}^4$ view to the space-time in which we sit.

Since $\mathbb{R}^1$ contains no angles, from where are derived the $90°$ angles with which copies of $\mathbb{R}^1$ are fitted together? The reader might respond that she observes the angles, but she observes them in only 4-dimensional space-time and cannot *a priori* be sure that 4-dimensional space-time is of the form $\mathbb{R}^4$. Further, the reader observes an angle (a velocity boost) between the time axis and a spatial axis that is of a very different type from the type of angle she observes between two spatial axes. Why two different types of angle? Surely, this means two types of space. Why not three or four types of angle, or, in a single space, a single type of angle? It does seem sensible that a single space should have only one type of angle.

It seems that we have experience of only one type of space. This is the 4-dimensional space-time in which we sit. No-one has ever seen 3-dimensional space; by this it is meant that there is no observational evidence that 3-dimensional space is a sub-space of 4-dimensional space-time. Although it is generally, unthinkingly, assumed to be so, no-one has ever stopped the flow of time to check this. Nor has anyone ever presented to the world a space of two spatial dimensions and one time dimension in which the third spatial dimension is of zero thickness. No-one has ever seen a 2-dimensional space separate from our 4-dimensional space-time, although we can argue that it does exist and that we know what it is because we have the 2-dimensional rotation matrices. No-one has ever seen 1-dimensional space[3] or 5-dimensional space. With thought, the reader will realise that the whole of humankind's physical understanding of empty space is based on observation of only one space, the 4-dimensional space-time in which we sit, and that we have little understanding of why this space-time is of the observed dimension and form.

If the nature of empty space is not $\mathbb{R}^n$, then what is its nature? What other types of space could exist? Perhaps at least part of the reason why the $\mathbb{R}^n$ view of space is so prevalent is that no-one has proposed an alternative view. However, there is an alternative view of the nature of space. That alternative view is what this book is about.

An alternative view of the nature of space:
The complex numbers, $\mathbb{C}$, correspond to positions in the 2-dimensional complex plane. The complex plane can be seen as a type of space. The complex plane has one real axis and one imaginary axis. Similarly, the quaternions, $\mathbb{H}$, correspond to positions in the 4-dimensional quaternion space which is comprised of one real axis and three imaginary axes. The view of space presented in this work is that n-dimensional space is of the form of one real axis and $(n-1)$ imaginary axes.

The basis of this idea is that one should be able to rotate within a geometric space. This necessitates a rotation matrix. By rotation matrix, we do not mean the 'false' rotation matrices that have zeros or ones as some of their elements and which form the $SO(p,q)$ Lie groups. A 'true' n-dimensional rotation matrix has a n-dimensional

---

[3] We can take the view that the real numbers are 1-dimensional space.

trigonometric function as every one of its elements. Examples of such 'true' rotation matrices are the two 2-dimensional rotation matrices:

$$\begin{bmatrix} \cos\theta & \sin\theta \\ -\sin\theta & \cos\theta \end{bmatrix} \quad \& \quad \begin{bmatrix} \cosh\chi & \sinh\chi \\ \sinh\chi & \cosh\chi \end{bmatrix} \qquad (1.1)$$

Examples of 'false' rotation matrices are:

$$\begin{bmatrix} \cos\theta & \sin\theta & 0 \\ -\sin\theta & \cos\theta & 0 \\ 0 & 0 & 1 \end{bmatrix} \quad \& \quad \begin{bmatrix} \cos\theta & 0 & \sin\theta \\ 0 & 1 & 0 \\ -\sin\theta & 0 & \cos\theta \end{bmatrix} \qquad (1.2)$$

We hold the view that these 'false' 3-dimensional rotation matrices are just 2-dimensional rotation matrices dressed up as 3-dimensional rotation matrices.

A rotation matrix multiplied by a real number (homothetic matrix of the appropriate size) is the polar form of a division algebra. It follows that any space in which there is rotation must also be a division algebra. In the case of the two 2-dimensional rotation matrices above, (1.1), the division algebras are the complex numbers, $\mathbb{C}$, and the hyperbolic complex numbers, $\mathbb{S}$, respectively.

"Hang on," your author hears the reader cry. "There are no 3-dimensional complex numbers nor any 4-dimensional complex numbers with the $t^2 - x^2 - y^2 - z^2$ norm that is the distance function of the space-time in which we sit." The reader is in error to think that there are no 3-dimensional complex numbers; they are not widely known, but we will shortly reveal them. There are many types of complex numbers in all dimensions. There are two types of 2-dimensional complex numbers. There are four types of 3-dimensional complex numbers (two of them are algebraically isomorphic). There are twenty-four types of 4-dimensional complex numbers (many of them are algebraically isomorphic). There are sixteen types of 5-dimensional complex numbers. Although we are at present unable to easily calculate[4] exactly how many types of n-

---

[4] We can do the calculation, but it is not easy for higher dimensions.

dimensional complex numbers there are for any value of n, we do know that for any prime number, p, there are 2^{p-1} types of p-dimensional complex number (many of which will be algebraically isomorphic). We also know exactly how to find the mathematical expression of such n-dimensional complex numbers for any given n; it is easy, but, for higher n, it can be a cumbersome calculation.

The reader would be correct to say there are no 4-dimensional complex numbers with the $t^2 - x^2 - y^2 - z^2$ norm. This work will present the view that the space-time in which we sit is not a single space but is an aggregation of six spaces – that is a sum of six (algebraically isomorphic) types of complex numbers. More accurately, we will argue that the space-time in which we sit is the sum of one algebra in six different bases. In the introduction to this work, we mentioned that the non-compatibility of the angular momentum operators seems to necessitate three different bases in our space. We will see in due course that the mathematics predicts six 4-dimensional A_3 spaces (three spaces and three anti-spaces), together with two 4-dimensional electromagnetic spaces (electromagnetic space and anti-electromagnetic space), and that the sum of the distances through all three of the A_3 spaces, or the three anti-spaces, or all six spaces, produces the form of the $t^2 - x^2 - y^2 - z^2$ norm with which we are familiar. This concept is quite shocking, but it is not really very different from calculating an expectation value. In a sense, we calculate the expectation distance function. We are saying that there are many different spaces sharing the same axes. This view also explains why 'microscopic physics' is quantitised but 'macroscopic physics' is not quantitised; each of the separate spaces has commutation relations; the sum of the separate spaces, since it is not a division algebra, does not have a valid type of multiplication and hence no valid commutation relations.

An alternative form of Lie algebra:

Conventional Lie groups, roughly, are 'spherical[5] surfaces' associated with rotation in the different conventional types of linear space $\{\mathbb{R}^n, \mathbb{C}^n, \mathbb{H}^n, \mathbb{O}^n\}$. Since we are adopting a different view of empty space, we will find different 'spherical surfaces' within these different types of space. Each of these different 'spherical surfaces' corresponds to a rotation matrix. Associated with the complex numbers, $\mathbb{C}$, there is a 2×2 rotation matrix containing the $\{\cos(\ ), \sin(\ )\}$ trigonometric functions. The set of points mapped by this rotation matrix, the unit circle, is isomorphic to the conventional Lie group $U(1)$. Associated with the non-commutative quaternions, there is a 4×4 rotation matrix containing the four quaternion trigonometric functions. This rotation matrix is isomorphic[6] to the conventional Lie group $SU(2)$. However, most rotation matrices are not isomorphic to conventional Lie groups. There is a $n \times n$ rotation matrix containing n different n-dimensional trigonometric functions associated with every type of n-dimensional complex number; it is no more than the angular part of the polar form of the algebra. There is thus a 'spherical surface' defined by each of these rotation matrices. It seems that, as we unveil higher dimensional complex numbers, we find new types of Lie groups to accompany our new types of space. Actually, we do a little more than this because the new type of 'Lie group theory' contains commutative groups (commutative 'spherical surfaces') as well as the non-commutative groups. In fact, these rotation matrices are spinors, but more on that later.

An alternative approach to space:

The existence of rotation matrices leads to a change of view regarding how we define a geometric space. Mathematicians normally define a geometric space by an invented distance function. Although it is often

---

[5] This is spherical in its broadest sense.
[6] Isomorphic in this sense means that the commutation relations are the same.

required that a distance function will satisfy the three metric space axioms, there is an unlimited number of such distance functions. We can adopt an approach different from this; we can define a geometric space by a rotation matrix rather than by a distance function. Surely, if a space is a geometric space in any meaningful sense of the phrase, then it should have angles and we should be able to rotate within that space and so there should be a rotation matrix associated with that space. A rotation matrix together with a real radial axis is a division algebra. We are thus, as we said above, associating the polar form of a type of complex numbers with a type of geometric space.

One cannot simply invent rotation matrices; they have to be discovered, and there are not many of them[7]. Defining a geometric space by a rotation matrix rather than by a distance function leads to only a small and finite number of types of space in any given dimension. There are only two types of 2-dimensional space and only four types of 3-dimensional space etc... This approach also leads to a much tighter mathematical structure for the space. The divergence, gradient, and curl of a vector field in the space are automatically defined as are the distance function and the type of symmetries in the space. Each rotation matrix corresponds to a particular, usually unconventional, Lie group. There are even particular tensors, like the electromagnetic tensor, automatically defined for the sum of the algebraically isomorphic spaces. There is no room for the mathematician to invent mathematics within such spaces in the way that mathematicians invent distance functions. The mathematics of such spaces is immutably set in stone; it will not be invented; it waits to be discovered.

There is much mathematics in this alternative view of space that is not to be found in the conventional view of space. In particular, finite groups play a central role in the understanding of empty space that is presented here; each type of space is underlain by a particular finite group of the same order as the dimension of the space. Trigonometry becomes generalised and expanded to include a particular set of trigonometric functions for each type of space. In non-commutative

---

[7] There are an infinite number of them, but it is a small infinite number.

spaces, there is a non-commutative inner product. Non-commutativity can be measured and used to define a geodesic through space (the space that is the Lie group). Deviation from the geodesic can be associated with force.

Chapter 2

Finite Groups and Division Algebras

We will be taking the view that a geometric space is the same thing as a division algebra. We therefore need to know a little about division algebras. For technical reasons, we state that we are concerned with only division algebras of zero characteristic; we have no interest in modular algebras. Zero characteristic means the numbers are continuous like the real numbers, $\mathbb{R}$, or the complex numbers, $\mathbb{C}$, and not discontinuous like 'clock arithmetic'.

The conventional view is that the only zero characteristic division algebras over the real numbers are the real numbers, $\mathbb{R}$, the complex numbers, $\mathbb{C}$, and the non-commutative (sometimes not included) quaternions, $\mathbb{H}$; depending upon taste, we do or do not count the non-associative 8-dimensional octonians, $\mathbb{O}$. In the conventional view, the commutative hyperbolic complex numbers, $\mathbb{S}$, first discovered by Cockle in 1848, and since then discovered independently over a dozen times by others, are usually unknown or ignored. They are not alone in this regard; there are an infinite number of zero characteristic division algebras both commutative and non-commutative almost all of which are unknown or ignored. They derive from the finite groups. We will be considering such zero characteristic division algebras, especially the non-commutative ones. Henceforward, we will not usually refer to the characteristic of a division algebra since we always deal with only zero characteristic algebras.

Division algebras and matrix notation:
The division algebra known as the complex numbers is often written as:

$$\mathbb{C} = a + ib \qquad (2.1)$$

12

However, this, as it stands, is an incomplete description. To make it complete, we need to add the side relation $i^2 = -1$. Without this side relation, the expression $a + ib$ is not an algebra (it is not of multiplicatively closed form). We have:

$$\mathbb{C} = a + ib$$
$$i^2 = -1$$

(2.2)

Similarly the quaternions are usually written as:

$$\mathbb{H} = a + \hat{i}b + jc + kd$$ (2.3)

This too is incomplete. The quaternions are properly written as:

$$\mathbb{H} = a + \hat{i}b + jc + kd$$
$$\hat{i}^2 = j^2 = k^2 = -1$$
$$\hat{i}j = k, \quad \hat{i}k = -j, \quad jk = \hat{i},$$
$$j\hat{i} = -k, \quad k\hat{i} = j, \quad kj = -\hat{i},$$

(2.4)

We need all these side relations to form an algebra. An alternative way of writing these algebras is as matrices.

$$\mathbb{C} = \begin{bmatrix} a & b \\ -b & a \end{bmatrix}, \qquad \mathbb{H} = \begin{bmatrix} a & b & c & d \\ -b & a & -d & c \\ -c & d & a & -b \\ -d & -c & b & a \end{bmatrix}$$

(2.5)

The matrices automatically include all the side relations which are required to form a division algebra. Henceforward we will adopt the matrix notation for division algebras[8].

Matrix multiplication (linear multiplication if you prefer) is the underlying multiplication of all (zero characteristic) division algebras. Traditionally, it is said that the multiplication of $\mathbb{C}$ is to do

---

[8] The change of notation brings much more than just an easier way of writing the algebra, as will become clear. It is this change of notation that led your author to all the research presented in this book and similar research elsewhere.

with the square root of minus unity and that this is the imaginary unit, $i = \sqrt{-1}$, but this has historically been, at least sub-consciously, questioned by mathematicians. i is called an 'imaginary' number as an expression of the reluctance of mathematicians to accept this concept. With the matrix notation, we see:

$$\begin{bmatrix} 0 & 1 \\ -1 & 0 \end{bmatrix}^2 = \begin{bmatrix} -1 & 0 \\ 0 & -1 \end{bmatrix} \tag{2.6}$$

The right-most of these matrices, (2.6), is algebraically isomorphic to the real number -1.

In general, anti-symmetric $2^n \times 2^n$ matrices similar to the above are square roots of -1, and the corresponding symmetric matrices are square roots of $+1$. We have:

$$\begin{bmatrix} 0 & 1 \\ 1 & 0 \end{bmatrix}^2 = \begin{bmatrix} 1 & 0 \\ 0 & 1 \end{bmatrix} \tag{2.7}$$

In a pedantic sense, a division algebra is no more than a set of matrices that satisfies the division algebra axioms[9]. One of these axioms is multiplicative closure – the product matrix must be of the same form as the two factor matrices that formed it. For example:

$$\begin{bmatrix} a & b \\ -b & a \end{bmatrix}\begin{bmatrix} c & d \\ -d & c \end{bmatrix} = \begin{bmatrix} ac-bd & ad+bc \\ -(ad+bc) & ac-bd \end{bmatrix} \tag{2.8}$$

$$(a+ib)(c+id) = (ac-bd) + i(ad+bc)$$

Another of these axioms is that we must have a multiplicative identity. This means that the elements on the leading diagonal must be equal, as they are above, (2.8). If we can find sets of matrices that are of multiplicatively closed form and have the elements of the leading diagonal equal, we are well on the way to satisfying the

---

[9] Your author prefers to replace 'division algebra axioms' with 'observed properties of the real numbers'; these are the same things. Your author sees axiomatic declarations to be religious pontifications but observed properties to be scientific observations. Normally, only philosophers worry about the difference.

division algebra axioms. Once we have such a set of matrices, all we need to form a division algebra from them is that they be non-singular (multiplicative inverse and no zero divisors) and have an additive inverse. We will deal with both of these requirements shortly.

Scaling parameters:
The two 'bits (axes)' of the complex numbers, $\mathbb{C}$, are independent of each other. (We decline from using the term orthogonal because it is associated with transposition of matrices and such transposition is not an algebraic operation.) Since these 'bits (axes)' are independent of each other, there is no reason to presume that they scale equally against each other. We could draw a (squashed) complex plane using 10 centimetres to each real unit and 1 centimetre to each imaginary unit without making any mathematical mistake. We can, of course, as we do, arbitrarily set the scaling of the imaginary axis to be equal to the scaling of the real axis and thus set the scaling parameter to be unity, but doing this hides the existence of the scaling parameter. The complete written form of the complex numbers includes, not only the side relations outlined above, but also the scaling parameter. For the quaternions, there are three scaling parameters, one for each imaginary axis. The scaling parameters can be either positive or negative but not zero. There is no traditional notation that includes the scaling parameters. In matrix form, we have:

$$\mathbb{C}_\lambda = \begin{bmatrix} a & b \\ \lambda b & a \end{bmatrix} \quad : \ \lambda < 0 \qquad (2.9)$$

Examination of the division algebra axioms shows that, for examples:

$$\mathbb{C}_{\lambda=-5} = \begin{bmatrix} a & b \\ -5b & a \end{bmatrix} \quad \& \quad \mathbb{C}_{\lambda=-\frac{1}{4}} = \begin{bmatrix} a & b \\ -\frac{1}{4}b & a \end{bmatrix} \qquad (2.10)$$

are division algebras isomorphic to the 'standard' $\lambda = -1$ complex numbers. The norm of the scaled complex numbers is the determinant:

$$d^2 = a^2 + \left(\sqrt{\lambda}b\right)^2 \qquad (2.11)$$

and we see that the axis in the imaginary direction is scaled by $\sqrt{\lambda}$ with respect to the real axis. The quaternions are:

$$\begin{bmatrix} a & b & c & d \\[2mm] \alpha b & a & \dfrac{\alpha}{\varepsilon}d & \varepsilon c \\[4mm] \eta c & -\dfrac{\eta}{\varepsilon}d & a & -\varepsilon b \\[4mm] -\dfrac{\alpha\eta}{\varepsilon^2}d & \dfrac{\eta}{\varepsilon}c & -\dfrac{\alpha}{\varepsilon}b & a \end{bmatrix} \quad : \quad \alpha < 0,\ \varepsilon > 0,\ \eta < 0 \quad (2.12)$$

It will turn out that the scaling parameters seem to correspond to the physical constants of the universe. The scaling parameter of the complex numbers, $\mathbb{C}$, seemingly corresponds to the inverse of 'aitch-bar', $\hbar$. The scaling parameter on the d variable (bottom left-hand corner) of the above quaternion seemingly corresponds to the electromagnetic fine structure constant; the other quaternion scaling parameters seem to be the electron charge and the velocity of light.

Permutation matrices:

Division algebras are based upon finite groups. In this work, we write finite groups of order n as sets of $n \times n$ permutation matrices; the group operation is matrix multiplication. A permutation matrix is no more than a square matrix with zeros and ones as elements distributed in such a way that there is a single one in each row and a single one in each column. There is a one to one correspondence between the $n!$ $n \times n$ permutation matrices and the $n!$ permutations of n objects. Since a finite group is 'the same thing' as a closed set of permutations, we have the whole of finite group theory within permutation matrices. The identity of the order n group is the $n \times n$ permutation matrix which has a one for every element on the leading diagonal – the matrix identity. We give some examples. There is only one order 2 finite group. It is the group C_2, and it is represented by the two permutation matrices:

$$C_2 \equiv \left\{ \begin{bmatrix} 1 & 0 \\ 0 & 1 \end{bmatrix} \& \begin{bmatrix} 0 & 1 \\ 1 & 0 \end{bmatrix} \right\} \qquad (2.13)$$

There is only one order 3 finite group. It is the group C_3, and it is represented by the three permutation matrices:

$$C_3 \equiv \left\{ \begin{bmatrix} 1 & 0 & 0 \\ 0 & 1 & 0 \\ 0 & 0 & 1 \end{bmatrix} \& \begin{bmatrix} 0 & 1 & 0 \\ 0 & 0 & 1 \\ 1 & 0 & 0 \end{bmatrix} \& \begin{bmatrix} 0 & 0 & 1 \\ 1 & 0 & 0 \\ 0 & 1 & 0 \end{bmatrix} \right\} \qquad (2.14)$$

There are two order 4 finite groups. One is the group C_4; the other is the group $C_2 \times C_2$, in which we will be particularly interested. The group $C_2 \times C_2$ is represented by the four permutation matrices:

$$C_2 \times C_2 \equiv \left\{ \begin{bmatrix} 1 & 0 & 0 & 0 \\ 0 & 1 & 0 & 0 \\ 0 & 0 & 1 & 0 \\ 0 & 0 & 0 & 1 \end{bmatrix} , \begin{bmatrix} 0 & 1 & 0 & 0 \\ 1 & 0 & 0 & 0 \\ 0 & 0 & 0 & 1 \\ 0 & 0 & 1 & 0 \end{bmatrix} \right. \\ \left. \begin{bmatrix} 0 & 0 & 1 & 0 \\ 0 & 0 & 0 & 1 \\ 1 & 0 & 0 & 0 \\ 0 & 1 & 0 & 0 \end{bmatrix} , \begin{bmatrix} 0 & 0 & 0 & 1 \\ 0 & 0 & 1 & 0 \\ 0 & 1 & 0 & 0 \\ 1 & 0 & 0 & 0 \end{bmatrix} \right\} \qquad (2.15)$$

The reader might notice that these $C_2 \times C_2$ matrices are symmetric matrices and hence have real eigenvalues and an orthogonal set of eigenvectors. These matrices have reflective symmetry across the leading diagonal and also across the opposing diagonal[10].

---

[10] The opposing diagonal runs from the top right-hand corner of the matrix to the bottom left-hand corner of the matrix.

Trivially, the 1×1 permutation matrix, $[1]$, represents the C_1 group.

A particular order n group is the set of $n \times n$ permutation matrices which 'fit together' in such a way that the sum of the permutation matrices is a matrix with a one for every element[11] – try it with (2.15) above. Mathematicians call this the regular representation of the group[12]. The essence, as far as this work is concerned, of the $n \times n$ permutation matrices that comprise an order n group is that the set of permutation matrices which form the group are multiplicatively closed; of course they are multiplicatively closed, they are a group; multiplicative closure is a group axiom. Multiplicative closure means that the product of any two of the matrices is another one of the matrices. We demonstrate with the C_3 group.

$$\begin{bmatrix} 0 & 1 & 0 \\ 0 & 0 & 1 \\ 1 & 0 & 0 \end{bmatrix}^2 = \begin{bmatrix} 0 & 0 & 1 \\ 1 & 0 & 0 \\ 0 & 1 & 0 \end{bmatrix} \qquad (2.16)$$

And:

$$\begin{bmatrix} 0 & 1 & 0 \\ 0 & 0 & 1 \\ 1 & 0 & 0 \end{bmatrix}\begin{bmatrix} 0 & 0 & 1 \\ 1 & 0 & 0 \\ 0 & 1 & 0 \end{bmatrix} = \begin{bmatrix} 1 & 0 & 0 \\ 0 & 1 & 0 \\ 0 & 0 & 1 \end{bmatrix} \qquad (2.17)$$

Division algebras:

We take each different permutation matrix of a group, and we turn the ones into a particular real variable. We do this in such a way that

---

[11] There are other ways of representing a group. For example, the order 6 group S_3 can be represented as six 3×3 permutation matrices or as six 6×6 permutation matrices. We prefer the 6×6 form.
[12] Howard Georgi: Lie Algebras in Particle Physics. Westview press: ISBN: 978-0-7382-0233-4

each permutation matrix has as elements the same variable in place of the ones. We demonstrate with C_2:

$$\begin{bmatrix} 1 & 0 \\ 0 & 1 \end{bmatrix} \rightarrow \begin{bmatrix} a & 0 \\ 0 & a \end{bmatrix}, \quad \begin{bmatrix} 0 & 1 \\ 1 & 0 \end{bmatrix} \rightarrow \begin{bmatrix} 0 & b \\ b & 0 \end{bmatrix} \tag{2.18}$$

We then add these matrices:

$$C_2 \rightarrow \begin{bmatrix} a & b \\ b & a \end{bmatrix} \tag{2.19}$$

We now have a matrix form which is of multiplicatively closed form[13]. By this we mean that the form of the matrix is maintained when we multiply two such matrices together:

$$\begin{bmatrix} a & b \\ b & a \end{bmatrix}\begin{bmatrix} c & d \\ d & c \end{bmatrix} = \begin{bmatrix} ac+bd & ad+bc \\ ad+bc & ac+bd \end{bmatrix} \tag{2.20}$$

We see that the two leading diagonal elements of the product are equal as is the case with the two factors and that the two off-diagonal elements of the product are equal as is the case with the two factors. This multiplicative closure of form is inherited from the multiplicative closure of the group. We see that we have the matrix form of the complex numbers, (2.9), except that the scaling parameter is +1 rather than −1. We still need non-singularity and an additive inverse.

Non-singularity:
The matrix form, (2.19), that we have derived from the group C_2 is not guaranteed non-singular. However, we can change it into something that is guaranteed non-singular by taking the exponential of the matrix:

---

[13] The matrix form is also the Cayley table of the group with all the identities on the leading diagonal, but this is unimportant. Such a presentation of the Cayley table is called the standard form of the Cayley table.

$$\exp\left(\begin{bmatrix} a & b \\ b & a \end{bmatrix}\right) = \begin{bmatrix} e^a & 0 \\ 0 & e^a \end{bmatrix}\begin{bmatrix} \cosh b & \sinh b \\ \sinh b & \cosh b \end{bmatrix} \quad (2.21)$$

This is the algebra that was first discovered by Cockle in 1848. It is most commonly known as the hyperbolic complex numbers, and it is the algebra of 2-dimensional space-time. It contains time dilation, length contraction, the relativistic velocity and acceleration transforms etc. Indeed, it deals with the 2-dimensional aspects of special relativity flawlessly[14], and it is therefore, in your author's opinion, better than Minkowski 4-vectors[15]. We have:

$$\begin{bmatrix} \cosh \chi & \sinh \chi \\ \sinh \chi & \cosh \chi \end{bmatrix} = \begin{bmatrix} \gamma & v\gamma \\ v\gamma & \gamma \end{bmatrix}$$

$$\gamma = \frac{1}{\sqrt{1-v^2}} \quad (2.22)$$

$$v = \tanh \chi$$

$$t' = t_0 \cosh \chi$$

Note that we have set $c = 1$ in the above (2.22). We've rewritten the usual time dilation formula above. So, we've hardly started, and already we have a good deal of physics falling out of the finite groups.

The additive inverse:

At this point, we need to review what we mean by division algebra. It is normally taken that a division algebra is a set of mathematical objects that satisfies the thirteen division algebra axioms. One of these axioms is the additive inverse axiom which necessitates the existence of negative real numbers. Looking at the hyperbolic complex numbers above, (2.21), we see there are no negative real numbers in this algebra. If we wish, the problem is easily fixed by forming the algebra as the union of two sets. One of these two sets is

[14] See Dennis Morris. Empty Space is Amazing Stuff. ISBN: 978-0-954-9780-7-5
[15] The norm of the acceleration 4-vector is imaginary – it ought not to be imaginary, and the conventional inner product of 4-vectors is questionable.

the set given above, (2.21), and the other is the set that is the negatives of these:

$$-\exp\left(\begin{bmatrix} a & b \\ b & a \end{bmatrix}\right) = \begin{bmatrix} -e^a & 0 \\ 0 & -e^a \end{bmatrix}\begin{bmatrix} \cosh b & \sinh b \\ \sinh b & \cosh b \end{bmatrix} \qquad (2.23)$$

If the reader wants to stick loyally to the axioms of a division algebra, then they will lose nothing by adding this negative set to the positive set, but it is a little messy. We prefer to dump the additive inverse axiom and do with only the derived positive set given above, (2.21). This might be a little shocking to the reader, but it is done with two thoughts in mind:

a) We did not get the negative set from the derivation given above.
b) We cannot travel backwards in time.

You see, the hyperbolic complex numbers correspond to 2-dimensional space-time. As they stand without the negative set, they correspond (2-dimensionally) to the space-time in which we find ourselves. With the negative set included, we allow movement backwards through time. We prefer observation over piety. Those loyal to the division algebra axioms might be correct. Possibly anti-matter is matter moving backwards in time and is thus associated with the negative set, but the mathematics which we reveal later can be interpreted to provide an alternative derivation of anti-matter, and so we proceed without the additive inverse axiom in this case. At our present state of knowledge, the dumping or keeping of the additive inverse axiom is no more than opinion, and your author's opinion might be wrong.

This view of ignoring the additive inverse is not as novel as it appears above. Within the Euclidean complex numbers, $\mathbb{C}$, we have the polar form:

$$e^{a+ib} = e^a\left(\cos b + i\sin b\right) \qquad (2.24)$$

In this, the additive inverse appears only because the Euclidean trigonometric functions, $\{\cos(\ ), \sin(\ )\}$, can both be negative. The additive inverse is absent from the hyperbolic complex numbers because the $\cosh(\ )$ function is never negative.

To put it another way; we are effectively defining a division algebra as a rotation matrix multiplied by a (positive) real number matrix. Such a definition does concur with the standard axioms except for the additive inverse axiom. (Of course, we need to define the object we call a rotation matrix.) A rotation matrix is exactly how we will define a geometric space, and thus it follows that a division algebra is the same thing as a geometric space. Actually, it is a spinor space, but more on that later.

How to derive forms of complex numbers:
Above, we have derived a division algebra from a finite group. The essential steps are:

1) Write the finite group of order n as a set of $n \times n$ permutation matrices.
2) Convert the ones in each permutation matrix to a particular real variable - using the same variable throughout a particular matrix.
3) Add the real variable matrices to form the algebraic matrix form.
4) Take the exponential of the algebraic matrix form to ensure non-singularity.

The resulting matrix will be a rotation matrix multiplied by a radial matrix. We say that this is a division algebra written in polar form – a type of complex number. The real variable of the basic algebraic matrix form, that is the one on the leading diagonal, will give the radial component of the polar form. The other variables, the imaginary variables, will form the rotation matrix of the polar form. The elements of the rotation matrix will be the n-dimensional trigonometric functions of the particular division algebra; there are n

of them. The angles that are the arguments of the trigonometric functions in the rotation matrix will be the $(n-1)$ imaginary variables, and so a 2-dimensional angle will be expressed as one argument, a 3-dimensional angle will be expressed as two arguments, and, in general, a n-dimensional angle will be expressed as $(n-1)$ arguments. For example, a 3-dimensional angle will be of the form (θ, ϕ) and a 4-dimensional angle will be of the form (θ, ϕ, φ). This view of an angle being an $(n-1)$-tuple of real numbers is quite a departure from our conventional view of an angle as being only a single real number. Of course, the $(n-1)$ arguments are orthogonal to each other, and we see that higher dimensional angles are vectors. The 2-dimensional angles to which we are accustomed are 1-dimensional vectors.

Scaling parameters again:
We have outlined above how to derive forms of complex numbers, but we have omitted a part of the process. We now address that omission. Above, we derived only one 2-dimensional division algebra from the group C_2. Although any non-trivial form of 2×2 matrix that is multiplicatively closed must necessarily be based on the group C_2, examination of the division algebra axioms shows that the C_2 algebraic matrix form we presented above is not the only form of 2×2 matrix that is multiplicatively closed. We have:

$$\begin{bmatrix} a & b \\ \lambda b & a \end{bmatrix} \begin{bmatrix} c & d \\ \lambda d & c \end{bmatrix} = \begin{bmatrix} ac + \lambda bd & ad + bc \\ \lambda(ad + bc) & ac + \lambda bd \end{bmatrix} \quad (2.25)$$

$$: \lambda \in \mathbb{R} \neq 0$$

The parameter, λ, is the scaling parameter that measures the relative scaling of the two axes. When $\lambda > 0$, we have the hyperbolic complex numbers, $\mathbb{S}_\lambda$. When $\lambda < 0$, we have the second 2-dimensional division algebra, $\mathbb{C}_\lambda$. When $\lambda = -1$, this is:

$$\exp\left(\begin{bmatrix} a & b \\ -b & a \end{bmatrix}\right) = \begin{bmatrix} e^a & 0 \\ 0 & e^a \end{bmatrix}\begin{bmatrix} \cos b & \sin b \\ -\sin b & \cos b \end{bmatrix} = \mathbb{C} \quad (2.26)$$

With which the reader is familiar.

Aside: In the case of the complex numbers, $\mathbb{C}$, there is no need to take the exponential of the algebraic matrix form since it is necessarily non-singular, but this is quite exceptional. We conjecture that the two 4-dimensional quaternion algebras which we will meet shortly are the only other algebras with this property.

In general, there are as many scaling parameters as there are imaginary axes, as one would expect if one does not think about it too deeply[16].

The distribution of the scaling parameters throughout an algebraic matrix form is not obvious and needs to be calculated. The calculation has the following steps:

1) We begin with the basic n-dimensional algebraic matrix form and we attach a potential scaling parameter to every element of that matrix.

2) We set the potential scaling parameters along the top row of the matrix equal to unity. We are effectively setting a basic scale, a normalisation if you prefer, for each variable in the algebra. This is the same as dividing each variable in the algebra by the potential scaling parameter of that variable on the top row of the matrix.

3) We set the potential scaling parameters of the leading diagonal equal. They must be equal if we are to have a multiplicative identity in the algebra, and all division algebras have such a multiplicative identity. We set them to unity to match the left-most element on the top row.

4) One by one, we eliminate the potential scaling parameters by examining the product of two basic matrices (with differently

---

[16] If one thinks about it deeply, it is rather miraculous that division algebras should exactly match empty space.

named variables but the same potential scaling parameters). We require the product matrix to be of the same form as the factor matrices; this is to require multiplicative closure. We require firstly the existence of a multiplicative identity (all leading diagonal elements are equal) and secondly multiplicative closure of the matrix form[17]. When we do this, we are conscious of the fact that the two matrices with potential scaling parameters are not usually commutative, but since we are seeking only multiplicative closure of form, this makes no difference to the results – see below (2.27) to (2.31)
.

5) We take the exponential of the matrix with the scaling parameters to get the division algebras.

It is remarkable, and by your author not clearly understood at a deep level, that insistence upon multiplicative closure of form leaves exactly $(n-1)$ scaling parameters – one for each imaginary axis.

We demonstrate the calculation technique with the C_3 algebras. From (2.14), we get the basic algebraic matrix form.

Stage 1:

$$
\begin{bmatrix} a & b & c \\ c & a & b \\ b & c & a \end{bmatrix} \rightarrow
\begin{bmatrix} \alpha a & \beta b & \chi c \\ \delta c & \varepsilon a & \phi b \\ \varphi b & \gamma c & \mu a \end{bmatrix}
\tag{2.27}
$$

Stage 2 & 3:

$$
\begin{bmatrix} \alpha a & \beta b & \chi c \\ \delta c & \varepsilon a & \phi b \\ \varphi b & \gamma c & \mu a \end{bmatrix} \rightarrow
\begin{bmatrix} a & b & c \\ \delta c & a & \phi b \\ \varphi b & \gamma c & a \end{bmatrix}
\tag{2.28}
$$

Stage 4:

---

[17] In fact, multiplicative closure of form necessitates equality of the leading diagonal elements. However, it is convenient to start with the leading diagonal.

$$\begin{bmatrix} a & b & c \\ \delta c & a & \phi b \\ \phi b & \gamma c & a \end{bmatrix}\begin{bmatrix} d & e & f \\ \delta f & d & \phi e \\ \phi e & \gamma f & d \end{bmatrix}$$

$$= \begin{bmatrix} ad + \delta bf + \phi ce & \sim & \sim \\ \sim & ad + \phi\gamma bf + \delta ce & \sim \\ \sim & \sim & ad + \phi bf + \phi\gamma ce \end{bmatrix} \quad (2.29)$$

$$\Rightarrow \phi = \delta, \quad \gamma = \frac{\delta}{\phi}$$

And:

$$\begin{bmatrix} d & e & f \\ \delta f & d & \phi e \\ \phi e & \gamma f & d \end{bmatrix}\begin{bmatrix} a & b & c \\ \delta c & a & \phi b \\ \phi b & \gamma c & a \end{bmatrix}$$

$$= \begin{bmatrix} ad + \phi bf + \delta ce & \sim & \sim \\ \sim & ad + \delta bf + \gamma\phi ce & \sim \\ \sim & \sim & ad + \phi\gamma bf + \phi ce \end{bmatrix} \quad (2.30)$$

$$\Rightarrow \phi = \delta, \quad \gamma = \frac{\delta}{\phi}$$

Leading to the algebraic form:

$$\exp\left(\begin{bmatrix} a & b & c \\ \delta c & a & \phi b \\ \delta b & \dfrac{\delta}{\phi}c & a \end{bmatrix}\right) \quad : \quad \{\delta, \phi\} \in \mathbb{R} \neq 0 \quad (2.31)$$

The calculation for higher dimensional algebras can be quite cumbersome.

The four different permutations of $\{\delta, \phi\} = \pm 1$ give the basic forms of the four C_3 algebras. By basic forms we mean the differing distributions of minus signs throughout the matrices. With the two scaling parameters set to the four permutations of ±1, those 3-dimensional algebras are:

$$\exp\left(\begin{bmatrix} a & b & c \\ c & a & b \\ b & c & a \end{bmatrix}\right), \quad \exp\left(\begin{bmatrix} a & b & c \\ c & a & -b \\ b & -c & a \end{bmatrix}\right)$$

$$\text{(2.32)}$$

$$\exp\left(\begin{bmatrix} a & b & c \\ -c & a & b \\ -b & -c & a \end{bmatrix}\right), \quad \exp\left(\begin{bmatrix} a & b & c \\ -c & a & -b \\ -b & c & a \end{bmatrix}\right)$$

Although we often set the scaling parameters to ±1, and the permutations of such values for the scaling parameters easily lead to the separate algebras, the scaling parameters must not be forgotten; they are part of the division algebras that are manifest in the physical interpretation of the mathematics. It would be more general to write the four C_3 algebras as:

$$\exp\left(\begin{bmatrix} a & b & c \\ \mu c & a & \kappa b \\ \mu b & \dfrac{\mu}{\kappa}c & a \end{bmatrix}\right), \quad \exp\left(\begin{bmatrix} a & b & c \\ \mu c & a & -\kappa b \\ \mu b & -\dfrac{\mu}{\kappa}c & a \end{bmatrix}\right)$$

$$\text{(2.33)}$$

$$\exp\left(\begin{bmatrix} a & b & c \\ -\mu c & a & \kappa b \\ -\mu b & -\dfrac{\mu}{\kappa}c & a \end{bmatrix}\right), \quad \exp\left(\begin{bmatrix} a & b & c \\ -\mu c & a & -\kappa b \\ -\mu b & \dfrac{\mu}{\kappa}c & a \end{bmatrix}\right)$$

$$\{\mu,\kappa\} > 0$$

It is not necessarily the case that the separate algebras that correspond to the different permutations of the scaling parameters are distinct. Most often, many of the algebras are algebraically isomorphic. In the case of the C_3 algebras derived above, only three of the four are algebraically non-isomorphic algebras. These three algebras have respectively, one cube root of minus unity and one cube root of plus unity (two algebras), two cube roots of minus unity, and two cube roots of plus unity[18]. The isomorphic algebras are the same algebra written in different bases.

When we come to interpret the mathematics into physics, we will see that it seems reality pays no heed to algebraic isomorphism. An example of such ignoring of algebraic isomorphism by reality is your author's two hands. My left hand is algebraically identical to my right hand, yet orientability exists in the physical world. Another example is the derivation of the electromagnetic tensor that we will see in a later chapter. Another example is the derivation of the distance function of the space-time in which we sit that we will see in a later chapter. If we have, say, six isomorphic algebras written in six different bases, reality uses all six of them. We cannot simply throw the seemingly superfluous algebras in the bin.

---

[18] In general, a cyclic group of order n will give rise to algebras with imaginary units that are the nth roots of plus or minus unity.

Chapter 3

Rotation Matrices, Trig. Funcs. and Distance Funcs.

We are primarily concerned with the polar form of an algebra because we see the rotation matrix as being, with the radial component, the algebra. The Cartesian form of an algebra is an algebra only if, to avoid singularity, we restrict the possible values of the variables to those which match the polar form. None-the-less, the Cartesian form is useful and we use it most often in calculations because it simplifies the calculation. A clear example is the familiar 2-dimensional complex numbers, $\mathbb{C}$:

$$\exp\left(\begin{bmatrix} a & b \\ -b & a \end{bmatrix}\right) = \begin{bmatrix} e^a & 0 \\ 0 & e^a \end{bmatrix} \begin{bmatrix} \cos b & \sin b \\ -\sin b & \cos b \end{bmatrix} \quad (3.1)$$

We justify the term rotation matrix by analogy with the established 2-dimensional rotation matrices. A rotation matrix is such that:

a) The product of two rotation matrices is a rotation matrix of the same nature. The trigonometric functions are conserved under matrix multiplication. An example of this is $\cos(\alpha)\cos(\beta) - \sin(\alpha)\sin(\beta) = \cos(\alpha+\beta)$. The positions of the trigonometric functions within the matrix are also conserved. This clearly works for commutative rotation matrices because, for commutative matrices $\{A, B\}$, we have $e^{(A+B)} = e^A e^B$. This also works for non-commutative rotation matrices, but we do not have the simple additive trigonometric relations of the commutative algebras.

b) The product of a rotation matrix with arguments $\{\alpha, \beta, \chi, ..\}$ and the same rotation matrix with arguments $\{-\alpha, -\beta, -\chi, ..\}$ is the identity matrix. This is closely connected to conjugation; actually, this is conjugation.

c) A rotation matrix contains functions called trigonometric functions that are projections from the unit n-dimensional sphere on to the axes of the space.

d) A rotation matrix is a linear transformation that preserves the distance from the origin – it has determinant unity. Hence, a product of two rotation matrices will have determinant unity.

3-dimensional trigonometric functions:

The reader is possibly unfamiliar with the 3-dimensional trigonometric functions[19] and the rotation matrix of the 3-dimensional algebras, and so we will briefly introduce these things to the reader. We have:

$$\exp\left(\begin{bmatrix} 0 & b & c \\ c & 0 & b \\ b & c & 0 \end{bmatrix}\right) = \begin{bmatrix} v_A(b,c) & v_B(b,c) & v_C(b,c) \\ v_C(b,c) & v_A(b,c) & v_B(b,c) \\ v_B(b,c) & v_C(b,c) & v_A(b,c) \end{bmatrix} \quad (3.2)$$

Where:

$$v_A(b,c) = \frac{1}{3}\left(e^{(b+c)} + 2e^{-\left(\frac{b+c}{2}\right)}\cos\left(\frac{\sqrt{3}}{2}(b-c)\right)\right) \quad (3.3)$$

$$v_B(b,c) = \frac{1}{3}\left(e^{(b+c)} + e^{-\left(\frac{b+c}{2}\right)}\left(\begin{array}{c} \sqrt{3}\sin\left(\frac{\sqrt{3}}{2}(b-c)\right) \\ -\cos\left(\frac{\sqrt{3}}{2}(b-c)\right) \end{array}\right)\right) \quad (3.4)$$

---

[19] For more details, see : Dennis Morris : Complex numbers The Higher Dimensional Forms

30

$$v_C(b,c) = \frac{1}{3}\left(e^{(b+c)} - e^{-\left(\frac{b+c}{2}\right)}\left(\sqrt{3}\sin\left(\frac{\sqrt{3}}{2}(b-c)\right) + \cos\left(\frac{\sqrt{3}}{2}(b-c)\right) \right) \right) \tag{3.5}$$

We justify the appellation 'trigonometric function' for these functions by emphasizing that they are each a projection on to an axis of the algebra from the 3-dimensional unit sphere[20] and that they appear in a rotation matrix. Since the polar form of an algebra corresponds to a (restricted) Cartesian form of that algebra, we have:

$$\begin{bmatrix} e^a & 0 & 0 \\ 0 & e^a & 0 \\ 0 & 0 & e^a \end{bmatrix}\begin{bmatrix} v_A(b,c) & v_B(b,c) & v_C(b,c) \\ v_C(b,c) & v_A(b,c) & v_B(b,c) \\ v_B(b,c) & v_C(b,c) & v_A(b,c) \end{bmatrix} = \begin{bmatrix} x & y & z \\ z & x & y \\ y & z & x \end{bmatrix} \tag{3.6}$$

Normalising both sides gives:

$$\begin{bmatrix} v_A(b,c) & v_B(b,c) & v_C(b,c) \\ v_C(b,c) & v_A(b,c) & v_B(b,c) \\ v_B(b,c) & v_C(b,c) & v_A(b,c) \end{bmatrix} = \frac{1}{e^a}\begin{bmatrix} x & y & z \\ z & x & y \\ y & z & x \end{bmatrix} \tag{3.7}$$

and we see that these functions are such projections from the unit sphere on to an axis.

These functions have series forms that are the matrix product of 3-way splittings of the exponential series[21]. In general, because a rotation matrix is basically the exponential of a sum of permutation matrices, n-dimensional trigonometric functions are closely related to the exponential series. We have:

---

[20] That is sphere in a very general sense that includes non-euclidean spheres.
[21] See : Dennis Morris: Complex Numbers The Higher Dimensional Forms

$$\exp\left(\begin{bmatrix} 0 & x & 0 \\ 0 & 0 & x \\ x & 0 & 0 \end{bmatrix}\right) = \begin{bmatrix} 1+\dfrac{x^3}{3!}+\dfrac{x^6}{6!}.. & \sim & \sim \\[2mm] \dfrac{x^2}{2!}+\dfrac{x^5}{5!}+\dfrac{x^8}{8!}... & \sim & \sim \\[2mm] x+\dfrac{x^4}{4!}+\dfrac{x^7}{7!}+... & \sim & \sim \end{bmatrix} \qquad (3.8)$$

These functions also have trigonometric relations that are analogous to the more familiar 2-dimensional trigonometric functions. Such relations arise as the product of two rotation matrices or as the determinant of the rotation matrix. Since, in general, the determinant of the exponential of a matrix with zero trace is unity, we have:

$$\det\left(\begin{bmatrix} v_A(b,c) & v_B(b,c) & v_C(b,c) \\ v_C(b,c) & v_A(b,c) & v_B(b,c) \\ v_B(b,c) & v_C(b,c) & v_A(b,c) \end{bmatrix}\right) = 1$$

$$v_A{}^3 + v_B{}^3 + v_C{}^3 - 3v_Av_Bv_C = 1 \qquad (3.9)$$

as direct calculation will verify. This is analogous to:

$$\det\left(\begin{bmatrix} \cos b & \sin b \\ -\sin b & \cos b \end{bmatrix}\right) = 1$$

$$\cos^2 b + \sin^2 b = 1 \qquad (3.10)$$

The existence of the trigonometric functions of each division algebra is part of the justification for considering a geometric space to be isomorphic to a division algebra rather than of the form $\mathbb{R}^n$. Trigonometric functions are geometric things.

Higher dimensional angles are vectors:

We note, as briefly mentioned above[22], that the 3-dimensional trigonometric functions have two arguments, (b,c). The 2-

---

[22] This concept is so shocking, that we introduce it slowly.

dimensional trigonometric functions have only one argument, (b). In general, a n-dimensional trigonometric function will have $(n-1)$ arguments; n-dimensional angles are comprised of $(n-1)$ variables. We are used to the idea that an angle is a scalar, and it is in 2-dimensional space, but higher dimensional angles are vectors. They are vectors within a spherical surface unit distance from the origin within the division algebra space[23]. This spherical surface is of dimension $(n-1)$, of course. Because higher dimensional angles are vectors, they have a direction.

In the case of non-commutative spaces, we will see that normally non-commutative rotations are commutative when they are through vector angles that are in the same direction. Successive rotations through vector angles that point in the same direction on a spherical surface are movements along a great circle (a geodesic). Thus, commutativity of two rotations is associated with movement along a geodesic and non-commutativity of two rotations is associated with deviation from a geodesic. Deviation from a geodesic is generally associated with force.

The 3-dimensional rotations shown above, (3.2), are all commutative, and so deviation from a geodesic is not possible in 3-dimensional space and so no concept of force can be associated with such deviation in 3-dimensional space. It seems that there is no such thing as 3-dimensional forces, and so one would expect that 3-dimensional space would not interact with anything. The same applies to all commutative spaces. It would seem that, except for the 2-dimensional spaces, the entire observable universe is associated with non-commutative spaces[24].

Perhaps the reader should pause to ponder the profundity expressed in the previous two paragraphs.

---

[23] This spherical surface is a group of infinite order whose group operation is rotation – a kind of Lie group.

[24] We seem to have slipped a profound idea into the main body of the text without mentioning the profundity of that idea.

Aside: A 1-dimensional complex number is a real number. The polar form of a real number is $e^{[a]} e^{[0]}$. The rotation matrix part of this is $e^{[0]} = 1$, and so we see that the 1-dimensional trigonometric function is just 1, which has zero arguments. Of course, 1 is the projection from the unit circle in 1-dimensional space on to the real axis.

Trigonometric addition relations:

Within the 2-dimensional spaces, we derive the trigonometric addition relations from a product of two rotation matrices. We demonstrate with the Euclidean complex numbers, $\mathbb{C}$. We have:

$$\exp\left(\begin{bmatrix} 0 & b \\ -b & 0 \end{bmatrix}\right) \exp\left(\begin{bmatrix} 0 & d \\ -d & 0 \end{bmatrix}\right) = \exp\left(\begin{bmatrix} 0 & b+d \\ -(b+d) & 0 \end{bmatrix}\right) \quad (3.11)$$

The above relationship is true only if the matrices are commutative. This relationship leads directly to:

$$\begin{bmatrix} \cos(b) & \sin(b) \\ -\sin(b) & \cos(b) \end{bmatrix} \begin{bmatrix} \cos(d) & \sin(d) \\ -\sin(d) & \cos(d) \end{bmatrix}$$
$$= \begin{bmatrix} \cos(b+d) & \sin(b+d) \\ -\sin(b+d) & \cos(b+d) \end{bmatrix}$$

$$(3.12)$$

$$\cos(b)\cos(d) - \sin(b)\sin(d) = \cos(b+d)$$
$$\cos(b)\sin(d) + \sin(b)\cos(d) = \sin(b+d)$$

The same can be done within any multiplicatively commutative algebra (it must be commutative) to get the addition relations of the trigonometric functions. We have relations like the 3-dimensional addition identity:

$$v_A(b+e, c+f) = v_A(b,c)v_A(e,f)$$
$$+ v_B(b,c)v_C(e,f) \quad (3.13)$$
$$+ v_C(b,c)v_B(e,f)$$

If the rotation matrices are non-commutative, we cannot derive such addition identities so easily and we have to deal with these matters differently (see later).

3-dimensional rotation is not 2-dimensional:

Within the 3-dimensional algebras (spaces), because C_3 does not have an order 2 sub-group, there is no 2-dimensional sub-algebra; this means there is no 2-dimensional sub-space[25]. Rotation in the 3-dimensional space is a 3-dimensional phenomenon; it does not happen in a 2-dimensional plane. Perhaps we should repeat that in case it slipped by the reader without shocking him.

Rotation in 3-dimensional space is not within a 2-dimensional plane.

The product of two '2-dimensional' 3-dimensional (sorry) matrices is not a '2-dimensional' matrix; the third element is not zero:

$$\begin{bmatrix} a & b & 0 \\ 0 & a & b \\ b & 0 & a \end{bmatrix}\begin{bmatrix} d & e & 0 \\ 0 & d & e \\ e & 0 & d \end{bmatrix} = \begin{bmatrix} ad & ae+bd & be \\ be & ad & ae+bd \\ ae+bd & be & ad \end{bmatrix} \quad (3.14)$$

$$\begin{bmatrix} v_A & 0 & v_C \\ v_C & v_A & 0 \\ 0 & v_C & v_A \end{bmatrix}\begin{bmatrix} v_A & 0 & v_C \\ v_C & v_A & 0 \\ 0 & v_C & v_A \end{bmatrix} = \begin{bmatrix} v_A^2 & v_C^2 & 2v_Av_C \\ 2v_Av_C & v_A^2 & v_C^2 \\ v_C^2 & 2v_Av_C & v_A^2 \end{bmatrix} \quad (3.15)$$

This is a general thing within division algebras. Rotation within the 5-dimensional division algebras is a 5-dimensional thing; it does not happen in a 2-dimensional plane or any plane other than a 5-dimensional one.

---

[25] This is a bit of a shocker when you first meet it, but, after 10 years of thinking about it every minute of every hour of every day, it becomes obvious.

4-dimensional rotation:

In 4-dimensional space, rotation is 4-dimensional, but, because the order four groups C_4 and $C_2 \times C_2$ have one and three order two sub-groups respectively, there is also rotation, in one or three respectively, 2-dimensional sub-spaces within these algebras. However, these are 4-dimensional 2-dimensional rotations (sorry again); they are different from the 2-dimensional rotations to which we are accustomed. For example, 4-dimensional 2-dimensional rotation in quaternion space (associated with the Lie group $SU(2)$) double covers[26] the 2-dimensional rotation in the Lie group $SO(3)$.

Within our space-time, we are accustomed to 2-dimensional rotation about an axis; for example, the eigenvalues and eigenvectors of the $SO(3)$ matrix:

$$\begin{bmatrix} \cos\theta & \sin\theta & 0 \\ -\sin\theta & \cos\theta & 0 \\ 0 & 0 & 1 \end{bmatrix} \tag{3.16}$$

include a constant eigenvector with eigenvalue 1:

$$\text{Eigenvector} \begin{bmatrix} 0 \\ 0 \\ 1 \end{bmatrix} \text{ with eigenvalue 1} \tag{3.17}$$

This eigenvector is unchanged by the rotation – it is an axis of rotation. However, the eigenvalues of the quaternion rotation matrix:

---

[26] Double cover means that we have twice as much rotation as we would expect. This is associated with the spin of the electron and the fact that its angular momentum is, ignoring quantum corrections, twice what we would expect. We sometimes see this expressed quite wrongly as the electron rotating through 720^0.

$$\begin{bmatrix} \cos\sqrt{\theta^2} & \sin\sqrt{\theta^2} & 0 & 0 \\ -\sin\sqrt{\theta^2} & \cos\sqrt{\theta^2} & 0 & 0 \\ 0 & 0 & \cos\sqrt{\theta^2} & -\sin\sqrt{\theta^2} \\ 0 & 0 & \sin\sqrt{\theta^2} & \cos\sqrt{\theta^2} \end{bmatrix} \tag{3.18}$$

Are of the form:

$$\cos\sqrt{\theta^2} \pm \sqrt{-1}\sin\sqrt{\theta^2} \tag{3.19}$$

These eigenvalues vary with the angle, and so this 2-dimensional 4-dimensional rotation is not rotation about an axis even through it is a 2-dimensional rotation in a higher dimensional space.

The *bona fide* 2-dimensional rotations occur in 2-dimensional spaces that are separate spaces in their own right; such 2-dimensional rotation does not affect any other dimension. We are familiar with such rotation. The 4-dimensional 2-dimensional rotations do effect the third and fourth dimensions. We compare a 4-dimensional quaternion 2-dimensional rotation and a 2-dimensional rotation:

$$\begin{bmatrix} \cos\theta & \sin\theta \\ -\sin\theta & \cos\theta \end{bmatrix} \begin{bmatrix} 0 & 0 & c & 0 \\ 0 & 0 & 0 & c \\ -c & 0 & 0 & 0 \\ 0 & -c & 0 & 0 \end{bmatrix} = ?! \tag{3.20}$$

This is meaningless.

However:

$$\begin{bmatrix} \cos\sqrt{\theta^2} & \sin\sqrt{\theta^2} & 0 & 0 \\ -\sin\sqrt{\theta^2} & \cos\sqrt{\theta^2} & 0 & 0 \\ 0 & 0 & \cos\sqrt{\theta^2} & -\sin\sqrt{\theta^2} \\ 0 & 0 & \sin\sqrt{\theta^2} & \cos\sqrt{\theta^2} \end{bmatrix} \begin{bmatrix} 0 & 0 & c & 0 \\ 0 & 0 & 0 & c \\ -c & 0 & 0 & 0 \\ 0 & -c & 0 & 0 \end{bmatrix}$$

$$\tag{3.21}$$

$$= \begin{bmatrix} 0 & 0 & c\cos\sqrt{\theta^2} & c\sin\sqrt{\theta^2} \\ 0 & 0 & -c\sin\sqrt{\theta^2} & c\cos\sqrt{\theta^2} \\ -c\cos\sqrt{\theta^2} & c\sin\sqrt{\theta^2} & 0 & 0 \\ -c\sin\sqrt{\theta^2} & -c\cos\sqrt{\theta^2} & 0 & 0 \end{bmatrix} \qquad (3.22)$$

We see that the 2-dimensional rotation matrix does not affect anything outside of the 2-dimensional space because we cannot multiply a 2×2 matrix by a 4×4 matrix. However the 4-dimensional 2-dimensional rotation does affect things that are outside of the 2-dimensional rotation plane.

In general, a quite important difference between 2-dimensional rotation and higher dimensional 2-dimensional rotation in the $C_2 \times C_2 \times...$ algebras is that higher dimensional 2-dimensional rotations are double covers of the 2-dimensional rotations[27]. I know I've just said that, but it is worth saying twice.

In due course, we will see that electromagnetism is a 4-dimensional quaternion phenomenon. However, when we derive the Lorentz force laws from the 4-potential, or, more easily, from the electromagnetic field, we do this by using the 2-dimensional rotation that is a Lorentz boost.

$$\begin{bmatrix} \cosh\chi & \sinh\chi \\ \sinh\chi & \cosh\chi \end{bmatrix}_{tx} \begin{bmatrix} E_z & B_y \\ B_y & E_z \end{bmatrix}$$

$$= \cosh_{tx}\chi \begin{bmatrix} E_z + v_x B_y & B_y + v_x E_z \\ B_y + v_x E_z & E_z + v_x B_y \end{bmatrix} \qquad (3.23)$$

$$= \gamma \begin{bmatrix} E_z + v_x B_y & B_y + v_x E_z \\ B_y + v_x E_z & E_z + v_x B_y \end{bmatrix}$$

---

[27] Double cover arises because the arguments of the $C_2 \times C_2 \times...$ algebras occur under a square root sign.

There are three similar calculations that lead to the Lorentz force laws[28]. If we used a 4-dimensional 2-dimensional rotation matrix to act upon the 4-dimensional 4-potential, we would not get the observed Lorentz force laws. This tells us something about the space-time in which we sit. It tells us that there are separate 2-dimensional spaces within this seemingly 4-dimensional space. Since 4-dimensional spaces do not have separate 2-dimensional sub-spaces, these 2-dimensional spaces (there are six of them) exist in their own right within the four dimensional space-time which we observe. We are therefore sitting in, at least, six separate spaces and not in only one space. So there's something to put in your pipe and cogitate upon.

Here is something else to cogitate upon; rotation in the complex plane, $\mathbb{C}$, is not rotation about an axis because there are only two dimensions in the complex plane.

Sub-algebras:

We have seen that, within a $C_2 \times C_2$ 4-dimensional rotation matrix, setting two imaginary variables to zero produces a 4-dimensional 2-dimensional rotation matrix and not a 2-dimensional rotation matrix within a separate 2-dimensional space. Similarly, within a 4-dimensional algebra, setting two imaginary variables to zero produces a 4-dimensional 2-dimensional sub-algebra and not a separate 2-dimensional sub-algebra. We are accustomed to the idea that the complex numbers, $\mathbb{C}$, are a sub-algebra of the quaternions, $\mathbb{H}$. In the traditional notation, this seems obviously true:

$$\mathbb{H} = a + \hat{i}b + jc + kd \qquad (3.24)$$

With $c = d = 0$:

$$\mathbb{H} = a + \hat{i}b \equiv ?\,\mathbb{C} \qquad (3.25)$$

Such notation is deceiving. The complex numbers, $\mathbb{C}$, are a separate 2-dimensional algebra in their own right. The quaternions with two

---

[28] We cover this in more detail later.

imaginary variables set to zero are a 4-dimensional 2-dimensional algebra that double covers the 2-dimensional complex numbers.

There might be physical consequences to this 'no separate sub-algebras' phenomenon. We will see in later chapters that electromagnetism is a 4-dimensional quaternion phenomenon. The 4-dimensional quaternion algebra does not exist as a separate sub-algebra in the 8-dimensional algebras. (It might exist in its 8-dimensional form.) We opine to associate higher dimensional algebras with higher energies. We see that electromagnetism does not exist as a separate phenomenon at higher energies. We would expect some kind of unification with higher energy phenomena.

Algebraic relations necessary for rotation:
When we multiply two rotation matrices together, we get a rotation matrix. This is another way of saying that the form of the matrices is maintained under multiplication. If the form of the matrices is maintained under multiplication, then the form of the determinant of the matrices is maintained under multiplication. Since, to every rotation matrix (the polar form of an algebra), there is a corresponding (normalised) Cartesian matrix (the restricted Cartesian form of the algebra), the form of the determinant of the Cartesian matrix is preserved under multiplication. In the 2-dimensional complex numbers, $\mathbb{C}$, we have:

$$\begin{bmatrix} a & b \\ -b & a \end{bmatrix}\begin{bmatrix} c & d \\ -d & c \end{bmatrix} = \begin{bmatrix} ac - bd & ad + bc \\ -(ad + bc) & ac - bd \end{bmatrix} \quad (3.26)$$

$$\det\left(\begin{bmatrix} a & b \\ -b & a \end{bmatrix}\right)\det\left(\begin{bmatrix} c & d \\ -d & c \end{bmatrix}\right) = \det\left(\begin{bmatrix} ac - bd & ad + bc \\ -(ad + bc) & ac - bd \end{bmatrix}\right) \quad (3.27)$$

$$\left(a^2 + b^2\right)\left(c^2 + d^2\right) = \left(ac - bd\right)^2 + \left(ad + bc\right)^2$$

This is an algebraic expression whose form is preserved under multiplication. That is:

$$\left(a^2 + b^2\right)\left(c^2 + d^2\right) = X^2 + Y^2 \qquad (3.28)$$

There must be an expression (the determinant of the restricted Cartesian form of the algebra) whose form is maintained under multiplication corresponding to every rotation matrix. A 3-dimensional example is the determinant of a 3-dimensional algebra.

$$\left(a^3 + b^3 + c^3 - 3abc\right)\left(d^3 + e^3 + f^3 - 3def\right)$$
$$= X^3 + Y^3 + Z^3 - 3XYZ \qquad (3.29)$$

There are four 3-dimensional division algebras, and so there are four such expressions. The other 3-dimensional (3-variable) expressions differ from the one above by only some sign changes.

Since there are 4-dimensional division algebras, there are 4-dimensional expressions whose form is maintained under multiplication; examples are:

$$\left(a^2 + b^2 + c^2 + d^2\right)\left(e^2 + f^2 + g^2 + h^2\right) = S^2 + T^2 + U^2 + V^2$$
$$\left(a^2 + b^2 - c^2 - d^2\right)\left(e^2 + f^2 - g^2 - h^2\right) = S^2 + T^2 - U^2 - V^2 \qquad (3.30)$$

The first of the above (3.30) is from[29] the determinant of the quaternions and the second is from the determinant of an algebra closely related to the quaternions and in which we are especially interested; we call it an A_3 algebra.

Aside: Clearly, for any prime, p, there are at least 2^{p-1} such p-dimensional expressions whose form is maintained under multiplication because there are 2^{p-1} p-dimensional division algebras. Some of the expressions will be effectively the same because some of the algebras will be isomorphic to each other. There are such expressions of any dimension.

---

[29] This expression is actually the square root of the quaternion determinant.

Distance functions:

The distance function or norm of a division algebra is the determinant of the Cartesian form of the algebra. We demonstrate with the hyperbolic complex numbers, $\mathbb{S}$. We have:

$$\det\left(\begin{bmatrix} r & 0 \\ 0 & r \end{bmatrix}\begin{bmatrix} \cosh\chi & \sinh\chi \\ \sinh\chi & \cosh\chi \end{bmatrix}\right) = \det\left(\begin{bmatrix} t & z \\ z & t \end{bmatrix}\right) \qquad (3.31)$$

$$r^2 = t^2 - z^2$$

No 4-dimensional rotation in our space-time:

The observed distance function of the 4-dimensional space-time in which we sit is of the form:

$$d^2 = t^2 - x^2 - y^2 - z^2 \qquad (3.32)$$

It is a simple algebraic fact that the form of this expression is not maintained under multiplication. Therefore, there cannot be a 4-dimensional rotation matrix associated with the space-time in which we sit. Therefore, the space-time in which we sit is not a division algebra space. It is easily observed that we cannot rotate 4-dimensionally in the space-time in which we sit. If we try to adopt the simple understanding of a 5-year old, we will say that, "A space is not a geometric space if it has no angles within it". Without a rotation matrix, there are no angles. Therefore, in the eyes of the 5-year old, the 4-dimensional space-time in which we sit is not a proper space in its own right.

Aside: We draw the reader's attention to the imbalance in the numbers of signs in the distance function of the space-time in which we sit; it is $3:1$ and not $2:2$. We will derive this 'unbalanced' distance function from the 'balanced' distance functions of the A_3 algebras in due course.

An A_3 rotation matrix:

We conclude this chapter with a preview of what is to come. We present a rotation matrix from one of the 4-dimensional division A_3 algebras that are derived from the $C_2 \times C_2$ group. Although the A_3 algebras are non-commutative, a particular matrix commutes with itself. Provided we calculate the exponential of the matrix all at once and do not separately calculate the exponentials of the individual imaginary variables and then take their product, we can calculate the exponential of a non-commutative matrix. We have set the scaling parameters to ± 1. An A_3 rotation matrix is:

$$A_3^{ROT} = \begin{bmatrix} \cosh(\lambda) & \dfrac{b}{\alpha}\sinh(\alpha) & \dfrac{c}{\alpha}\sinh(\alpha) & \dfrac{d}{\alpha}\sinh(\alpha) \\[2mm] -\dfrac{b}{\alpha}\sinh(\alpha) & \cosh(\alpha) & -\dfrac{d}{\alpha}\sinh(\alpha) & \dfrac{c}{\alpha}\sinh(\alpha) \\[2mm] \dfrac{c}{\alpha}\sinh(\alpha) & -\dfrac{d}{\alpha}\sinh(\alpha) & \cosh(\alpha) & -\dfrac{b}{\alpha}\sinh(\alpha) \\[2mm] \dfrac{d}{\alpha}\sinh(\alpha) & \dfrac{c}{\alpha}\sinh(\alpha) & \dfrac{b}{\alpha}\sinh(\alpha) & \cosh(\alpha) \end{bmatrix} \quad (3.33)$$

$$\alpha = \sqrt{-b^2 + c^2 + d^2}$$

Of course, being 4-dimensional, the trigonometric functions have three arguments, (b,c,d).

The astute reader will notice that this 4-dimensional rotation matrix reduces to three 4-dimensional 2-dimensional sub-rotation matrices if any two of the variables are zero. For example:

$$\begin{bmatrix} \cosh(\alpha) & 0 & \sqrt{+1}\sinh(\alpha) & 0 \\ 0 & \cosh(\alpha) & 0 & \sqrt{+1}\sinh(\alpha) \\ \sqrt{+1}\sinh(\alpha) & 0 & \cosh(\alpha) & 0 \\ 0 & \sqrt{+1}\sinh(\alpha) & 0 & \cosh(\alpha) \end{bmatrix} \quad (3.34)$$

$$\alpha = \sqrt{c^2}$$

These are, of course, 4-dimensional 2-dimensional rotations. When they act upon a 4-dimensional matrix (say a 4-dimensional potential), they affect all (t, x, y, z) components of that matrix and not only two of those components.

The more astute reader will notice that two of the 2-dimensional sub-rotations are space-time boosts but that the third is a spatial rotation. We have:

$$\begin{bmatrix} \cosh(\alpha) & \dfrac{b}{\alpha}\sinh(\alpha) & 0 & 0 \\[2ex] -\dfrac{b}{\alpha}\sinh(\alpha) & \cosh(\alpha) & 0 & 0 \\[2ex] 0 & 0 & \cosh(\alpha) & -\dfrac{b}{\alpha}\sinh(\alpha) \\[2ex] 0 & 0 & \dfrac{b}{\alpha}\sinh(\alpha) & \cosh(\alpha) \end{bmatrix}$$

$$\alpha = \sqrt{-b^2} = i\sqrt{b^2}$$

$$(3.35)$$

Using the established relations[30]:

$$\cosh(ix) = \cos(x)$$
$$\sinh(ix) = i\sin(x)$$

$$(3.36)$$

The above matrix (3.35) becomes:

---

[30] These relations are established in 2-dimensional space. We are in 4-dimensional space, and so we are making a technical error here to use them. The error is easily corrected; we replace the i with the square root of minus unity, $\sqrt{-1}$.

44

$$
\begin{bmatrix}
\cos\left(\sqrt{b^2}\right) & \sqrt{+1}\sin\left(\sqrt{b^2}\right) & 0 & 0 \\
-\sqrt{+1}\sin\left(\sqrt{b^2}\right) & \cos\left(\sqrt{b^2}\right) & 0 & 0 \\
0 & 0 & \cos\left(\sqrt{b^2}\right) & -\sqrt{+1}\sin\left(\sqrt{b^2}\right) \\
0 & 0 & \sqrt{+1}\sin\left(\sqrt{b^2}\right) & \cos\left(\sqrt{b^2}\right)
\end{bmatrix}
$$

(3.37)

The extremely astute reader will realise that the product of the $\{c,d\}$ variables is the b variable. We demonstrate with the Cartesian form:

$$
\begin{bmatrix}
0 & 0 & c & 0 \\
0 & 0 & 0 & c \\
c & 0 & 0 & 0 \\
0 & c & 0 & 0
\end{bmatrix}
\begin{bmatrix}
0 & 0 & 0 & d \\
0 & 0 & -d & 0 \\
0 & -d & 0 & 0 \\
d & 0 & 0 & 0
\end{bmatrix}
=
\begin{bmatrix}
0 & -cd & 0 & 0 \\
cd & 0 & 0 & 0 \\
0 & 0 & 0 & cd \\
0 & 0 & -cd & 0
\end{bmatrix}
$$

(3.38)

And:

$$
\begin{bmatrix}
0 & 0 & 0 & d \\
0 & 0 & -d & 0 \\
0 & -d & 0 & 0 \\
d & 0 & 0 & 0
\end{bmatrix}
\begin{bmatrix}
0 & 0 & c & 0 \\
0 & 0 & 0 & c \\
c & 0 & 0 & 0 \\
0 & c & 0 & 0
\end{bmatrix}
=
\begin{bmatrix}
0 & cd & 0 & 0 \\
-cd & 0 & 0 & 0 \\
0 & 0 & 0 & -cd \\
0 & 0 & cd & 0
\end{bmatrix}
$$

(3.39)

The commutator of two matrices is the difference of the products of those two matrices. If the two matrices are $\{A,B\}$, the commutator is written as $[A,B]$ and we have:

$$[A,B]=AB-BA \tag{3.40}$$

Above, (3.39), we have that the commutator of two space-time boosts is a spatial rotation. The reader will be aware that this is a basic property of the Lie group $SO(3,1)$ more commonly known as the Lorentz group which is intimately connected to the Pauli matrices.

Double cover:

Looking at (3.37), we see that inserting the angle (b) into the rotation matrix will produce the same rotation as inserting the angle $(-b)$. This is what is called double cover; it is very different from the 2-dimensional trigonometric functions that do not have a square root over the angles.

Groups within groups (infinite order):

In the previous chapters, we have derived rotation matrices from finite groups. Every finite group has rotation matrices within it; we simply take the exponential of the matrix sum of the non-identity elements (variables in permutation matrices) and this is a rotation matrix. However, a rotation matrix is a group in its own right. We allow the angle variables to vary from $-\infty \to \infty$ (or sometimes a lesser interval), we get all the points on a spherical surface of unit distance from the origin. This set of points is a group of infinite order. We are close to doing Lie algebra (see later).

This 'groups within groups' is easier to visualise in the 3-dimensional space with which we are familiar. The 3-dimensional space is a group of infinite order in that any movement (rotation or translation or both) will go from a point in the space to a point in the space. A 2-dimensional spherical surface within this 3-dimensional space is also a group of infinite order in that any rotation will go from a point in the spherical surface to a point in the spherical surface. Any great circle in the spherical surface is also a group of infinite order.

Chapter 4

Inner Products in Division Algebras

A vector in a division algebra has one real component and $(n-1)$ imaginary components. A vector in a division algebra is no more than a number of that algebra, but we still have curls, divergence and gradient. A vector field in the complex plane, $\mathbb{C}$, is of the form:

$$\begin{bmatrix} f(x,y) \\ g(x,y) \end{bmatrix} \equiv \begin{bmatrix} f(x,y) & g(x,y) \\ -g(x,y) & f(x,y) \end{bmatrix} \tag{4.1}$$

A particular vector in the complex plane $\mathbb{C}$ is of the form:

$$\begin{bmatrix} a & b \\ -b & a \end{bmatrix} \tag{4.2}$$

Inner product and outer product:
At this stage, we deal with inner products within only commutative algebras. We deal with inner products of non-commutative algebras later. The inner product of two vectors in a commutative algebra comes connected to the outer product of those vectors. These products are given by the product of the conjugate of one normalised vector with the other normalised vector. We call this the angle product and denote it by $\odot$:

$$\vec{u} \odot \vec{v} = \frac{1}{\sqrt{a^2+b^2}} \begin{bmatrix} a & b \\ -b & a \end{bmatrix} \odot \frac{1}{\sqrt{c^2+d^2}} \begin{bmatrix} c & d \\ -d & c \end{bmatrix} \tag{4.3}$$

$$= \frac{1}{\sqrt{a^2+b^2}\sqrt{c^2+d^2}} \begin{bmatrix} a & -b \\ b & a \end{bmatrix} \begin{bmatrix} c & d \\ -d & c \end{bmatrix} \tag{4.4}$$

$$= \frac{1}{\sqrt{a^2+b^2}\sqrt{c^2+d^2}} \begin{bmatrix} ac+bd & ad-bc \\ -(ad-bc) & ac+bd \end{bmatrix}$$

$$= \begin{bmatrix} \vec{u} \bullet \vec{v} & \vec{u} \times \vec{v} \\ -\vec{u} \times \vec{v} & \vec{u} \bullet \vec{v} \end{bmatrix}$$

(4.5)

Aside: The reader might recognise this as the Clifford product within the Clifford algebra $Cl_{2,0}$: $\vec{e_1}\vec{e_1} = \vec{e_2}\vec{e_2} = +1$:

$$\left(a\vec{e_1} + b\vec{e_2}\right)\left(c\vec{e_1} + d\vec{e_2}\right) = ac + bd + (ad - bc)\vec{e_{12}}$$

(4.6)

We also have:

$$\vec{v} \odot \vec{u} = \frac{1}{\sqrt{c^2+d^2}} \begin{bmatrix} c & d \\ -d & c \end{bmatrix} \odot \frac{1}{\sqrt{a^2+b^2}} \begin{bmatrix} a & b \\ -b & a \end{bmatrix}$$

$$= \frac{1}{\sqrt{c^2+d^2}\sqrt{a^2+b^2}} \begin{bmatrix} c & -d \\ d & c \end{bmatrix} \begin{bmatrix} a & b \\ -b & a \end{bmatrix}$$

$$= \frac{1}{\sqrt{c^2+d^2}\sqrt{a^2+b^2}} \begin{bmatrix} ac+bd & -(ad-bc) \\ ad-bc & ac+bd \end{bmatrix}$$

$$= \begin{bmatrix} \vec{u} \bullet \vec{v} & -\vec{u} \times \vec{v} \\ \vec{u} \times \vec{v} & \vec{u} \bullet \vec{v} \end{bmatrix}$$

(4.7)

The cross product within the $\mathbb{C}$ algebra is not a vector that protrudes from the complex plane at 90^0 to that plane. The complex plane is a 2-dimensional space; it cannot 'grow' another dimension. The interpretation of the cross product as a vector perpendicular to the plane of the two vectors does not apply within division algebras. (It does not apply in any spaces other than $\mathbb{R}^3$ & $\mathbb{R}^7$, and it is questionable in those spaces.) It is, of course, useful to 3-dimensional engineers in spite of its non-algebraic usage.

If we take the angle product of the two vectors in polar form, we get:

$$\vec{u} \odot \vec{v} = \begin{bmatrix} r_1 & 0 \\ 0 & r_1 \end{bmatrix} \begin{bmatrix} \cos\theta & \sin\theta \\ -\sin\theta & \cos\theta \end{bmatrix} \odot \begin{bmatrix} r_2 & 0 \\ 0 & r_2 \end{bmatrix} \begin{bmatrix} \cos\phi & \sin\phi \\ -\sin\phi & \cos\phi \end{bmatrix}$$

$$= \begin{bmatrix} r_1 r_2 & 0 \\ 0 & r_1 r_2 \end{bmatrix} \begin{bmatrix} \cos\theta & -\sin\theta \\ \sin\theta & \cos\theta \end{bmatrix} \begin{bmatrix} \cos\phi & \sin\phi \\ -\sin\phi & \cos\phi \end{bmatrix} \qquad (4.8)$$

$$= \begin{bmatrix} r_1 r_2 & 0 \\ 0 & r_1 r_2 \end{bmatrix} \begin{bmatrix} \cos(\theta-\phi) & \sin(\theta-\phi) \\ -\sin(\theta-\phi) & \cos(\theta-\phi) \end{bmatrix}$$

Since the polar form is equal to the Cartesian form, we have, after normalisation of both forms:

$$\begin{bmatrix} \cos(\theta-\phi) & \sin(\theta-\phi) \\ -\sin(\theta-\phi) & \cos(\theta-\phi) \end{bmatrix}$$

$$= \frac{1}{\sqrt{a^2+b^2}\sqrt{c^2+d^2}} \begin{bmatrix} ac+bd & ad-bc \\ -(ad-bc) & ac+bd \end{bmatrix} \qquad (4.9)$$

Which is:

$$\cos(\theta-\phi) = \frac{ac+bd}{\sqrt{a^2+b^2}\sqrt{c^2+d^2}}$$

$$\sin(\theta-\phi) = \frac{ad-bc}{\sqrt{a^2+b^2}\sqrt{c^2+d^2}} \qquad (4.10)$$

This is the familiar 'the cosine of the angle between two vectors is the dot product of the two vectors' result and the corresponding result for the sine of the angle and the magnitude of the cross product. Within a commutative division algebra, the inner product of two complex numbers (vectors) is no more than a measure of the angle between the two complex numbers; in general, this measure is expressed as the leading trigonometric function of the space[31]. Of course, every division algebra has its own type of angle, but, within a commutative algebra, the inner product is still a measure of that angle. (For non-commutative angles, the measure of the angle is

---

[31] The leading trigonometric function is the one that projects on to the real axis and appears on the leading diagonal of the rotation matrix.

slightly more complicated.) The inner product of two 3-dimensional vectors is the 3-dimensional angle between those vectors. We have:

$$v_A(b-e,c-f)$$

$$= \frac{a^2d + b^2e + c^2f - abf - ace - bcd}{\left(\sqrt[3]{a^3 + b^3 + c^3 - 3abc}\right)^2 \sqrt[3]{d^3 + e^3 + f^3 - 3def}} \tag{4.11}$$

There is a tendency among mathematicians to think of the inner product as something that they impose on to a metric space[32] to give that space geometric structure. In such cases, the form of the inner product is copied from the invented distance function and is almost invariably set equal to the cosine function. We have seen above that the cosine function exists within only the complex plane, $\mathbb{C}$; it does not exist outside of this 2-dimensional algebra[33]. As such, the cosine function is associated with only the distance function of the complex plane which is $d^2 = x^2 + y^2$. Perhaps it is nonsensical to associate this 2-dimensional trigonometric function with any other distance function[34].

The commutative inner product measures the angle between two vectors. The angle between two vectors and the length (norm) of a vector are invariant under change of co-ordinate system. This is why physicists are interested in inner products.

---

[32] A metric space is defined by an invented distance function.

[33] Something very similar to the cosine function does 'accidently' appear in the 4-dimensional algebras, but the 4-dimensional version of the cosine function is a different thing from the 2-dimensional cosine function; it is a double cover of the 2-dimensional cosine.

[34] Of course, the 4-dimensional distance function $d^2 = t^2 + x^2 + y^2 + z^2$ contains the 2-dimensional distance function $d^2 = x^2 + y^2$.

Chapter 5

Differentiation in Division Algebras

At this stage, we deal with differentiation within only commutative division algebras. We will deal with non-commutative differentiation later when we meet the quaternions. We demonstrate using the complex numbers, $\mathbb{C}$, but this technique is applicable to all commutative division algebras; with appropriate modification, it is applicable to non-commutative division algebras. Again, for ease of presentation, we use the Cartesian form of the algebra[35].

The differential of the function, vector field, potential, or whatever:

$$\vec{A} = \begin{bmatrix} f(x,y) & g(x,y) \\ -g(x,y) & f(x,y) \end{bmatrix} \tag{5.1}$$

With respect to:

$$\begin{bmatrix} x & y \\ -y & x \end{bmatrix} \tag{5.2}$$

is:

$$\frac{\partial \begin{bmatrix} f(x,y) & g(x,y) \\ -g(x,y) & f(x,y) \end{bmatrix}}{\partial \begin{bmatrix} x & y \\ -y & x \end{bmatrix}} \tag{5.3}$$

---

[35] We reiterate that for most division algebras, the Cartesian form is not an algebra; only the polar form is an algebra. Of course, in the above example, we use $\mathbb{C}$ whose Cartesian form is an algebra.

$$
= \frac{\partial \begin{bmatrix} f(x,y) & g(x,y) \\ -g(x,y) & f(x,y) \end{bmatrix}}{\partial \begin{bmatrix} x & 0 \\ 0 & x \end{bmatrix}} + \frac{\partial \begin{bmatrix} f(x,y) & g(x,y) \\ -g(x,y) & f(x,y) \end{bmatrix}}{\partial \begin{bmatrix} 0 & y \\ -y & 0 \end{bmatrix}} \tag{5.4}
$$

This is:

$$
= \frac{\partial \begin{bmatrix} f(x,y) & 0 \\ 0 & f(x,y) \end{bmatrix}}{\partial \begin{bmatrix} x & 0 \\ 0 & x \end{bmatrix}} + \frac{\partial \begin{bmatrix} 0 & g(x,y) \\ -g(x,y) & 0 \end{bmatrix}}{\partial \begin{bmatrix} x & 0 \\ 0 & x \end{bmatrix}}
$$

$$
+ \frac{\partial \begin{bmatrix} f(x,y) & 0 \\ 0 & f(x,y) \end{bmatrix}}{\partial \begin{bmatrix} 0 & y \\ -y & 0 \end{bmatrix}} + \frac{\partial \begin{bmatrix} 0 & g(x,y) \\ -g(x,y) & 0 \end{bmatrix}}{\partial \begin{bmatrix} 0 & y \\ -y & 0 \end{bmatrix}} \tag{5.5}
$$

The left-most of the above four differentials (5.5) is simply a real number function differentiated with respect to a real number. It is of the form:

$$
\frac{\partial \begin{bmatrix} f(x,y) & 0 \\ 0 & f(x,y) \end{bmatrix}}{\partial \begin{bmatrix} x & 0 \\ 0 & x \end{bmatrix}} = \begin{bmatrix} \dfrac{\partial f}{\partial x} & 0 \\ 0 & \dfrac{\partial f}{\partial x} \end{bmatrix} \tag{5.6}
$$

The second left-most of the above four differentials, (5.5), is an imaginary function differentiated with respect to a real variable. It is of the form:

$$
\frac{\partial \begin{bmatrix} 0 & g(x,y) \\ -g(x,y) & 0 \end{bmatrix}}{\partial \begin{bmatrix} x & 0 \\ 0 & x \end{bmatrix}} = \begin{bmatrix} 0 & 1 \\ -1 & 0 \end{bmatrix} \frac{\partial \begin{bmatrix} g(x,y) & 0 \\ 0 & g(x,y) \end{bmatrix}}{\partial \begin{bmatrix} x & 0 \\ 0 & x \end{bmatrix}} \tag{5.7}
$$

$$= \begin{bmatrix} 0 & 1 \\ -1 & 0 \end{bmatrix} \begin{bmatrix} \dfrac{\partial g}{\partial x} & 0 \\ 0 & \dfrac{\partial g}{\partial x} \end{bmatrix}$$

$$= \begin{bmatrix} 0 & \dfrac{\partial g}{\partial x} \\ -\dfrac{\partial g}{\partial x} & 0 \end{bmatrix}$$

(5.8)

The third left-most of the above four differentials, (5.5), is a real function differentiated with respect to an imaginary variable. It is of the form:

$$\frac{\partial \begin{bmatrix} f(x,y) & 0 \\ 0 & f(x,y) \end{bmatrix}}{\partial \begin{bmatrix} 0 & y \\ -y & 0 \end{bmatrix}} = \frac{1}{\begin{bmatrix} 0 & 1 \\ -1 & 0 \end{bmatrix}} \frac{\partial \begin{bmatrix} f(x,y) & 0 \\ 0 & f(x,y) \end{bmatrix}}{\partial \begin{bmatrix} y & 0 \\ 0 & y \end{bmatrix}}$$

$$= \begin{bmatrix} 0 & -1 \\ 1 & 0 \end{bmatrix} \begin{bmatrix} \dfrac{\partial f}{\partial y} & 0 \\ 0 & \dfrac{\partial f}{\partial y} \end{bmatrix}$$

(5.9)

$$= \begin{bmatrix} 0 & -\dfrac{\partial f}{\partial y} \\ \dfrac{\partial f}{\partial y} & 0 \end{bmatrix}$$

The right-most of the above four differentials, (5.5), is an imaginary function differentiated with respect to an imaginary variable. It is of the form:

$$\frac{\partial \begin{bmatrix} 0 & g(x,y) \\ -g(x,y) & 0 \end{bmatrix}}{\partial \begin{bmatrix} 0 & y \\ -y & 0 \end{bmatrix}} = \frac{\begin{bmatrix} 0 & 1 \\ -1 & 0 \end{bmatrix} \partial \begin{bmatrix} g(x,y) & 0 \\ 0 & g(x,y) \end{bmatrix}}{\partial \begin{bmatrix} y & 0 \\ 0 & y \end{bmatrix}}$$

(5.10)

$$= \begin{bmatrix} \dfrac{\partial g}{\partial y} & 0 \\[2mm] 0 & \dfrac{\partial g}{\partial y} \end{bmatrix} \tag{5.11}$$

The complete differential is the sum of the four separate differentials, (5.6) to (5.11):

$$\frac{\partial \begin{bmatrix} f(x,y) & g(x,y) \\ -g(x,y) & f(x,y) \end{bmatrix}}{\partial \begin{bmatrix} x & y \\ -y & x \end{bmatrix}} = \begin{bmatrix} \dfrac{\partial f}{\partial x} + \dfrac{\partial g}{\partial y} & \dfrac{\partial g}{\partial x} - \dfrac{\partial f}{\partial y} \\[3mm] -\left(\dfrac{\partial g}{\partial x} - \dfrac{\partial f}{\partial y}\right) & \dfrac{\partial f}{\partial x} + \dfrac{\partial g}{\partial y} \end{bmatrix} \tag{5.12}$$

Thinking of the function as a vector field, $\vec{A} = \begin{bmatrix} f(x,y) \\ g(x,y) \end{bmatrix}$, the reader will recognise this differential as:

$$\frac{\partial \begin{bmatrix} f(x,y) & g(x,y) \\ -g(x,y) & f(x,y) \end{bmatrix}}{\partial \begin{bmatrix} x & y \\ -y & x \end{bmatrix}} = \begin{bmatrix} Div\left(\vec{A}\right) & Curl\left(\vec{A}\right) \\[2mm] -Curl\left(\vec{A}\right) & Div\left(\vec{A}\right) \end{bmatrix} \tag{5.13}$$

If we had differentiated a scalar field (with $g(x,y) = 0$), we would have:

$$\frac{\partial \begin{bmatrix} f(x,y) & 0 \\ 0 & f(x,y) \end{bmatrix}}{\partial \begin{bmatrix} x & y \\ -y & x \end{bmatrix}} = \begin{bmatrix} \dfrac{\partial f}{\partial x} & -\dfrac{\partial f}{\partial y} \\[3mm] \dfrac{\partial f}{\partial y} & \dfrac{\partial f}{\partial x} \end{bmatrix} = grad\left(\vec{A}\right) \tag{5.14}$$

Notice that the minus sign has swapped position. This concurs with the view that the gradient of a vector field is a 1-form (covariant vector). Within division algebras, the conjugate of the complex number (vector) is the dual form of the vector. We have:

$$\begin{bmatrix} a & -b \\ b & a \end{bmatrix} \begin{bmatrix} a & b \\ -b & a \end{bmatrix} = \begin{bmatrix} a^2 + b^2 & 0 \\ 0 & a^2 + b^2 \end{bmatrix} \in \mathbb{R} \qquad (5.15)$$

Further differentiation will give the Laplacian of a vector field.

The technique of matrix differentiation is an essential tool within division algebra mathematics. The technique is simple but cumbersome. It can be made less cumbersome, and with this more confusing, by using differential operators like the momentum operator of quantum mechanics which is the $U(1)$ differential operator or the angular momentum operator of quantum mechanics which is the non-commutative $SU(2)$ differential operator. Although it is not made obvious in standard quantum mechanics texts, a differential operator is no more than a calculation shortcut that obfuscates the mathematics.

2-dimensional hyperbolic vector fields:

Above, we have differentiated the complex numbers, $\mathbb{C}$, and come to the gradient, divergence, and curl within the complex plane. Obviously, we can differentiate the hyperbolic complex numbers, $\mathbb{S}$, and obtain a gradient, divergence, and curl within the 2-dimensional space-time that is the complex plane of these numbers. Again, for ease, we use the Cartesian form of the algebra but remind the reader that only the polar form is an algebra. We have:

$$\frac{\partial \begin{bmatrix} f(t,z) & g(t,z) \\ g(t,z) & f(t,z) \end{bmatrix}}{\partial \begin{bmatrix} t & z \\ z & t \end{bmatrix}} = \begin{bmatrix} \dfrac{\partial f}{\partial t} + \dfrac{\partial g}{\partial z} & \dfrac{\partial g}{\partial t} + \dfrac{\partial f}{\partial z} \\ \dfrac{\partial g}{\partial t} + \dfrac{\partial f}{\partial z} & \dfrac{\partial f}{\partial} + \dfrac{\partial g}{\partial z} \end{bmatrix} \qquad (5.16)$$

Thus, we have the space-time gradient as:

$$grad\left(\overrightarrow{ST}\right) = \begin{bmatrix} \dfrac{\partial f}{\partial t} & \dfrac{\partial f}{\partial z} \\ \dfrac{\partial f}{\partial z} & \dfrac{\partial f}{\partial t} \end{bmatrix} \qquad (5.17)$$

And the space-time divergence and curl as:

$$\begin{bmatrix} Div\left(\overrightarrow{ST}\right) & Curl\left(\overrightarrow{ST}\right) \\ Curl\left(\overrightarrow{ST}\right) & Div\left(\overrightarrow{ST}\right) \end{bmatrix} = \begin{bmatrix} \dfrac{\partial f}{\partial t} + \dfrac{\partial g}{\partial z} & \dfrac{\partial g}{\partial t} + \dfrac{\partial f}{\partial z} \\ \dfrac{\partial g}{\partial t} + \dfrac{\partial f}{\partial z} & \dfrac{\partial f}{\partial} + \dfrac{\partial g}{\partial z} \end{bmatrix} \qquad (5.18)$$

We see that the space-time curl is the sum of two differentials rather than the difference of two differentials. Normally, we associate the curl with a force.

Within normal vector calculus, we consider only one type of curl. That is because we work in only one type of space. Each type of space has its own type of gradient, its own type of divergence[36], and its own type of curl. We really ought to add a subscript to the gradient, divergence, and curl of each space to indicate what we are dealing with:

$$Grad_{\mathbb{C}}, \; Div_{\mathbb{C}}, \; Curl_{\mathbb{C}}$$
$$Grad_{\mathbb{S}}, \; Div_{\mathbb{S}}, \; Curl_{\mathbb{S}} \qquad (5.19)$$

Well, perhaps we didn't really ought to add these subscripts after all; they look messy.

In the differentiation above, we habitually took the constant matrices out of the differential to the left. Since the complex numbers, $\mathbb{C}$ & $\mathbb{S}$, are both multiplicatively commutative division algebras (algebraic fields), it makes no difference whether we take the constant matrices to the left or to the right of the differential. When we come to non-commutative division algebras like the quaternions, we will be conscious of which way, left or right, we extract the

---

[36] If a divergence is zero, we have the conservation of some kind of charge, and so, perhaps, we have a kind of charge associated with each type of space.

constant matrices from the differentials. Except for that, which is not a small thing, differentiation within any division algebra can be done using the technique outlined above.

Differential operators:
We can view differentiation as an operator. We simply matrix multiply the differential operator by the vector field. The differential operator in $\mathbb{C}$ is:

$$\begin{bmatrix} \dfrac{\partial}{\partial x} & -\dfrac{\partial}{\partial y} \\ \dfrac{\partial}{\partial y} & \dfrac{\partial}{\partial x} \end{bmatrix} \tag{5.20}$$

Notice the position of the minus sign. Within the detailed version of the differentiation, for example (5.9), we had to take the inverse of the imaginary variable:

$$\begin{bmatrix} 0 & 1 \\ -1 & 0 \end{bmatrix}^{-1} = \begin{bmatrix} 0 & -1 \\ 1 & 0 \end{bmatrix} \tag{5.21}$$

For anti-symmetric imaginary variables, the inverse is the conjugate, but this is not the case with symmetric imaginary variables which, being the square roots of plus one, are their own inverses. The $\mathbb{S}$ differential operator is:

$$\begin{bmatrix} \dfrac{\partial}{\partial t} & \dfrac{\partial}{\partial z} \\ \dfrac{\partial}{\partial z} & \dfrac{\partial}{\partial t} \end{bmatrix} \tag{5.22}$$

There is no minus sign.

Differentiation is then just the (kind of) matrix product of this operator with the vector field:

$$\begin{bmatrix} \dfrac{\partial}{\partial x} & -\dfrac{\partial}{\partial y} \\[2ex] \dfrac{\partial}{\partial y} & \dfrac{\partial}{\partial x} \end{bmatrix} \begin{bmatrix} f(x,y) & g(x,y) \\[1ex] -g(,y) & f(x,y) \end{bmatrix} = \begin{bmatrix} \dfrac{\partial f}{\partial x}+\dfrac{\partial g}{\partial y} & \dfrac{\partial g}{\partial x}-\dfrac{\partial f}{\partial y} \\[2ex] -\left(\dfrac{\partial g}{\partial x}-\dfrac{\partial f}{\partial y}\right) & \dfrac{\partial f}{\partial x}+\dfrac{\partial g}{\partial y} \end{bmatrix} \quad (5.23)$$

Well! That's the differential operator. Personally, I think it is ugly and obfuscating. Methinks that mathematics and physics might be better off if we rid ourselves of such operators, but it does save time and paper.

Of course, we are free to differentiate with respect to only the real variable or only the imaginary variable, and such differentiations are themselves differential operators. Since the Lie group $U(1)$ is associated with rotation is the complex plane, we see there is an association between $U(1)$ and the $\mathbb{C}$ differential operator. We will later see that differentiation with respect to the imaginary variable within $\mathbb{C}$ is the quantum mechanical momentum operator. This is how we will deduce that the scaling parameter in the complex numbers, $\mathbb{C}$, is the inverse of the physical constant $\hbar$. In the hyperbolic complex numbers, $\mathbb{S}$, the scaling parameter corresponds to the limiting velocity of the universe. When we look at quaternions, we will find another (non-commutative) differential operator; in the quaternion case, the differential operator is associated with the Lie group $SU(2)$.

Chapter 6

Symmetries in Different Spaces

Within theoretical physics, we have the notion of symmetry. One such symmetry is rotational symmetry. We see that the physics of the universe is invariant under rotation in the 2-dimensional Euclidean plane. We see that the physics of the universe is invariant under rotation in 2-dimensional space-time – this is the theory of special relativity. We have seen above that different types of space have different types of rotation matrix with different trigonometric functions in them. We have the familiar 2-dimensional rotation in the Euclidean plane that we see as two dimensions in the 4-dimensional space-time in which we sit. We have the wholly unfamiliar 3-dimensional rotation in the 3-dimensional C_3 spaces. We have met 4-dimensional rotation above when we introduced an A_3 rotation matrix; we also met the 4-dimensional quaternion rotation matrix.

Aside: The quaternion rotation matrix is isomorphic to the Lie group $SU(2)$. In spite of their isomorphism, there are differences between $SU(2)$ and the quaternion rotation matrix. Unitary Lie groups are seen as being spherical surfaces based on rotations in $\mathbb{C}^n$ space. $SU(2)$ is based on rotation in $\mathbb{C}^2$ space. Quaternion rotation is rotation in quaternion space. In this work, we take the view that $\mathbb{C}^n$ space does not exist, and so, technically, we take the view that $SU(2)$ does not exist. The task before us is to rewrite theoretical physics using the quaternion rotation matrix in place of $SU(2)$; this is not a small task. Of course, the commutation relations that we usually call $SU(2)$ exist within the quaternions.

3-dimensional parity:

If different types of space have different types of rotation, then different types of space will have different types of rotational symmetry. If different spaces have different types of rotational symmetry, we might expect that different types of space will have different types of other symmetries; for example, we might expect different types of reflective symmetries. Within physics, we have the concept of parity. This being based upon the trigonometric functions $\{\cos(\ ), \sin(\ )\}$ which are seen as being even and odd functions. The cosine function is even because $\cos(\theta) = \cos(-\theta)$. The sine function is odd because $\sin(\theta) = -\sin(-\theta)$. Although physics does not often use the $\{\cosh(\ ), \sinh(\ )\}$ functions as even and odd functions, they are such functions.

Let us imagine that the Euclidean trigonometric functions were both even, and let us rotate first through the angle θ and then through the angle $-\theta$. Normally, such rotation followed by an equal reverse rotation would return us to the identity matrix, but when the trigonometric functions are both even, we have:

$$
\begin{bmatrix} \cos\theta & \sin\theta \\ -\sin\theta & \cos\theta \end{bmatrix} \begin{bmatrix} \cos(-\theta) & \sin(-\theta) \\ -\sin(-\theta) & \cos(-\theta) \end{bmatrix}
$$
$$
\overset{?}{=} \begin{bmatrix} \cos\theta & \sin\theta \\ -\sin\theta & \cos\theta \end{bmatrix} \begin{bmatrix} \cos(\theta) & \sin(\theta) \\ -\sin(\theta) & \cos(\theta) \end{bmatrix}
$$
$$
= \begin{bmatrix} \cos^2\theta - \sin^2\theta & 2\cos\theta\sin\theta \\ -2\cos\theta\sin\theta & \cos^2\theta - \sin^2\theta \end{bmatrix}
$$
$$
\neq \begin{bmatrix} 1 & 0 \\ 0 & 1 \end{bmatrix}
$$

(6.1)

Wherein we have used a '$=?$' equals sign to avoid asserting the equality of things that are not equal. For a rotation followed by an equal reverse rotation to return to where we started, we need the evenness/oddness of the trigonometric functions. We associate this evenness/oddness with reflective symmetry/anti-symmetry.

In 3-dimensions, we have:

$$PROD \begin{pmatrix} \begin{bmatrix} v_A(b,c) & v_B(b,c) & v_C(b,c) \\ v_C(b,c) & v_A(b,c) & v_B(b,c) \\ v_B(b,c) & v_C(b,c) & v_A(b,c) \end{bmatrix} \\ \begin{bmatrix} v_A(-b,-c) & v_B(-b,-c) & v_C(-b,-c) \\ v_C(-b,-c) & v_A(-b,-c) & v_B(-b,-c) \\ v_B(-b,-c) & v_C(-b,-c) & v_A(-b,-c) \end{bmatrix} \end{pmatrix} \qquad (6.2)$$

$$= \begin{bmatrix} 1 & 0 & 0 \\ 0 & 1 & 0 \\ 0 & 0 & 1 \end{bmatrix}$$

Aside: Looking at the 3-dimensional trigonometric functions, (3.5), and considering the complications of matrix multiplication, is it not remarkable that we get the identity matrix from multiplying such complicated things together in such a complicated way. It is such results that give humankind confidence in our endeavour to comprehend everything.

It seems that we must have some kind of evenness/oddness in this 3-dimensional space. The graphs of the 3-dimensional trigonometric functions should show that evenness/oddness. The graphs of these functions are 3-dimensional, but we can get the essence of them by taking a 2-dimensional slice through these 3-dimensional graphs. What we see is 3-dimensional evenness/oddness. This is parity in a 3-dimensional sense. The graphs are:

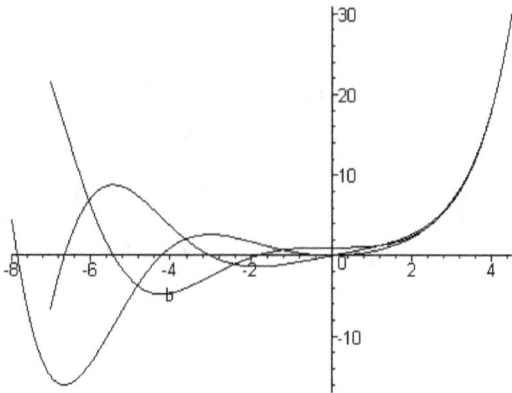

This is nothing like 2-dimensional parity. We see waves of exponentially increasing amplitude on the left-hand side and behaviour like the $\{\cosh(\ ),\sinh(\ )\}$ functions on the right-hand side.

Every type of space has its own type of symmetry:
The point is that every type of space has its own types of symmetry. Both the type of rotation and the parity (reflection) of the space seem to be inextricably connected to the trigonometric functions of the space.

If you go to a university physics department, you will see people walking about repeatedly saying "symmetry this" or "symmetry that" or just "symmetry, symmetry" and "What a symmetrical day it is" or "How symmetrical of you" or "Would you like a symmetrical cup of tea?" and "Do you take symmetrical sugar?" Within physics, symmetry is everywhere; we see above that we have barely dipped our toe into the ocean when it comes to understanding symmetry.

Chapter 7

4-dimensional Space

Having prepared the reader for a different view of space, we now begin to seek the space-time in which we sit. We will not complete our search in only one chapter, but we will make a start.

Our interest in this work is in non-commutative algebras because there seems to be within the universe a correlation between non-commutative variables (called generators[37] in Lie group theory) and forces. The standard model presumes there are eight gluons (strong force bosons) because there are eight non-commutative generators in the Lie group $SU(3)$. (This is in spite of the fact that there are only six things for these gluons to do.) The standard model also asserts that there are three weak force bosons because there are three non-commutative generators in $SU(2)$.

The 4-dimensional spaces are derived from the two order four groups $C_4 \& C_2 \times C_2$. The division algebras that derive from the C_4 group are all commutative. Since, it seems that commutative spaces do not interact with our universe in any way, they are of no interest to the physicist and we will not look at them in this work[38]. The other order four group $C_2 \times C_2$ is a commutative group, but, remarkably, it holds non-commutative division algebras (spaces) within it. We proceed in the established manner.

The permutation matrices of the $C_2 \times C_2$ group are:

---

[37] 'Generators of rotations' to give them their full title.
[38] See: Dennis Morris : Complex Numbers, the Higher dimensional forms

$$\left\{ \begin{bmatrix} 1 & 0 & 0 & 0 \\ 0 & 1 & 0 & 0 \\ 0 & 0 & 1 & 0 \\ 0 & 0 & 0 & 1 \end{bmatrix}, \begin{bmatrix} 0 & 1 & 0 & 0 \\ 1 & 0 & 0 & 0 \\ 0 & 0 & 0 & 1 \\ 0 & 0 & 1 & 0 \end{bmatrix}, \begin{bmatrix} 0 & 0 & 1 & 0 \\ 0 & 0 & 0 & 1 \\ 1 & 0 & 0 & 0 \\ 0 & 1 & 0 & 0 \end{bmatrix}, \begin{bmatrix} 0 & 0 & 0 & 1 \\ 0 & 0 & 1 & 0 \\ 0 & 1 & 0 & 0 \\ 1 & 0 & 0 & 0 \end{bmatrix} \right\} \tag{7.1}$$

We insert variables and sum these to form the basic Cartesian algebraic matrix form:

$$\begin{bmatrix} a & b & c & d \\ b & a & d & c \\ c & d & a & b \\ d & c & b & a \end{bmatrix} \tag{7.2}$$

We insert the potential scaling parameters (as Greek letters):

$$\begin{bmatrix} a & b & c & d \\ \alpha b & a & \delta d & \varepsilon c \\ \eta c & \theta d & a & \kappa b \\ \mu d & \nu c & \pi b & a \end{bmatrix} \tag{7.3}$$

and we seek to eliminate all but three of these potential scaling factors by requiring firstly a multiplicative identity and secondly multiplicative closure of the form. The matrix form with potential scaling parameters is non-commutative, however, leading diagonals of the product in either order are equal, and we can use these to eliminate some parameters without having to worry about non-commutativity. We will do this slowly because, in your author's opinion, it is a very beautiful piece of mathematics. We have the products:

$$AE = \begin{bmatrix} a & b & c & d \\ \alpha b & a & \delta d & \varepsilon c \\ \eta c & \theta d & a & \kappa b \\ \mu d & \nu c & \pi b & a \end{bmatrix} \begin{bmatrix} e & f & g & h \\ \alpha f & e & \delta h & \varepsilon g \\ \eta g & \theta h & e & \kappa f \\ \mu h & \nu g & \pi f & e \end{bmatrix} \tag{7.4}$$

$$EA = \begin{bmatrix} e & f & g & h \\ \alpha f & e & \delta h & \varepsilon g \\ \eta g & \theta h & e & \kappa f \\ \mu h & vg & \pi f & e \end{bmatrix} \begin{bmatrix} a & b & c & d \\ \alpha b & a & \delta d & \varepsilon c \\ \eta c & \theta d & a & \kappa b \\ \mu d & vc & \pi b & a \end{bmatrix} \quad (7.5)$$

The leading diagonal elements are:

$$\begin{aligned}
AE_{[1,1]} &= EA_{[1,1]} = ae + \alpha bf + \eta cg + \mu dh \\
AE_{[2,2]} &= EA_{[2,2]} = ae + \alpha bf + \varepsilon vcg + \delta\theta dh \\
AE_{[3,3]} &= EA_{[3,3]} = ae + \pi\kappa bf + \eta cg + \delta\theta dh \\
AE_{[4,4]} &= EA_{[4,4]} = ae + \pi\kappa bf + \varepsilon vcg + \mu dh
\end{aligned} \quad (7.6)$$

For a multiplicative identity, we need these four elements to be equal. From this we have:

$$\mu = \delta\theta, \quad v = \frac{\eta}{\varepsilon}, \quad \pi = \frac{\alpha}{\kappa} \quad (7.7)$$

We have eliminated three of the nine potential scaling parameters. It does not matter which particular parameters we choose to eliminate; the essence of the result is the same. The reader's attention is drawn to the fact that the three above equations (7.7) are all linear – they have only one solution.

Putting these solutions into the algebraic matrix form and taking the product gives:

$$AE := \begin{bmatrix} a & b & c & d \\ \alpha b & a & \delta d & \varepsilon c \\ \eta c & \theta d & a & \kappa b \\ \delta\theta d & \dfrac{\eta}{\varepsilon}c & \dfrac{\alpha}{\kappa}b & a \end{bmatrix} \begin{bmatrix} e & f & g & h \\ \alpha f & e & \delta h & \varepsilon g \\ \eta g & \theta h & e & \kappa f \\ \delta\theta h & \dfrac{\eta}{\varepsilon}g & \dfrac{\alpha}{\kappa}f & e \end{bmatrix} \quad (7.8)$$

$$EA := \begin{bmatrix} e & f & g & h \\ \alpha f & e & \delta h & \varepsilon g \\ \eta g & \theta h & e & \kappa f \\ \delta\theta h & \dfrac{\eta}{\varepsilon}g & \dfrac{\alpha}{\kappa}f & e \end{bmatrix} \begin{bmatrix} a & b & c & d \\ \alpha b & a & \delta d & \varepsilon c \\ \eta c & \theta d & a & \kappa b \\ \delta\theta d & \dfrac{\eta}{\varepsilon}c & \dfrac{\alpha}{\kappa}b & a \end{bmatrix} \quad (7.9)$$

Seeking multiplicative closure of form and comparing the elements $\{[2,1],[1,2]\}$ gives:

$$\alpha.AE_{[1,2]} = \alpha af + \alpha be + \alpha\theta ch + \frac{\alpha\eta}{\varepsilon}dg$$

$$AE_{[2,1]} = \alpha af + \alpha be + \delta\varepsilon\theta ch + \delta\eta dg$$

$$\alpha.EA_{[1,2]} = \alpha af + \alpha be + \frac{\alpha\eta}{\varepsilon}ch + \alpha\theta dg \qquad (7.10)$$

$$EA_{[2,1]} = \alpha af + \alpha be + \delta\eta ch + \delta\varepsilon\theta dg$$

We see that the non-commutativity does not affect the elimination of the next parameter. We have the linear equation:

$$\varepsilon = \frac{\alpha}{\delta} \qquad (7.11)$$

Seeking multiplicative closure of form and comparing the elements $\{[3,1],[1,3]\}$ gives:

$$\eta.AE_{[1,3]} = \eta ag + \eta ce + \eta\delta bh + \frac{\alpha\eta}{\kappa}df$$

$$AE_{[3,1]} = \eta ag + \eta ce + \delta\kappa\theta bh + \alpha\theta df$$

$$\eta.EA_{[1,3]} = \eta ag + \eta ce + \frac{\alpha\eta}{\kappa}bh + \eta\delta df \qquad (7.12)$$

$$EA_{[3,1]} = \eta ag + \eta ce + \alpha\theta bh + \delta\kappa\theta df$$

Again, the non-commutativity does not affect the elimination of the next parameter. We have the linear equation:

$$\kappa = \frac{\eta}{\theta} \tag{7.13}$$

Putting these into the algebraic matrix form and taking the product gives:

$$AE := \begin{bmatrix} a & b & c & d \\ \alpha b & a & \delta d & \dfrac{\alpha}{\delta}c \\ \eta c & \theta d & a & \dfrac{\eta}{\theta}b \\ \delta\theta d & \dfrac{\eta\delta}{\alpha}c & \dfrac{\alpha\theta}{\eta}b & a \end{bmatrix} \begin{bmatrix} e & f & g & h \\ \alpha f & e & \delta h & \dfrac{\alpha}{\delta}g \\ \eta g & \theta h & e & \dfrac{\eta}{\theta}f \\ \delta\theta h & \dfrac{\eta\delta}{\alpha}g & \dfrac{\alpha\theta}{\eta}f & e \end{bmatrix}$$

$$EA := \begin{bmatrix} e & f & g & h \\ \alpha f & e & \delta h & \dfrac{\alpha}{\delta}g \\ \eta g & \theta h & e & \dfrac{\eta}{\theta}f \\ \delta\theta h & \dfrac{\eta\delta}{\alpha}g & \dfrac{\alpha\theta}{\eta}f & e \end{bmatrix} \begin{bmatrix} a & b & c & d \\ \alpha b & a & \delta d & \dfrac{\alpha}{\delta}c \\ \eta c & \theta d & a & \dfrac{\eta}{\theta}b \\ \delta\theta d & \dfrac{\eta\delta}{\alpha}c & \dfrac{\alpha\theta}{\eta}b & a \end{bmatrix} \tag{7.14}$$

A non-linear equation:
We now discover that, regardless of non-commutativity and regardless of which elements we compare in quest of multiplicative closure, we are driven to:

$$\theta^2 = \frac{\eta^2 \delta^2}{\alpha^2} \tag{7.15}$$

This is not a linear equation. It has two solutions:

$$\theta = +\frac{\eta\delta}{\alpha} \quad \& \quad \theta = -\frac{\eta\delta}{\alpha} \tag{7.16}$$

It seems that the right-most of these two equations holds within it the whole of the 4-dimensional part of the observed universe; it certainly holds electromagnetism, Pauli matrices, some Lie algebras, and the space-time in which we sit; it also seems to hold gravity as expressed in general relativity.

It is most rare within the division algebras derived from finite groups to have a non-linear equation involved in the elimination of potential scaling parameters. Other than at a calculation level, this phenomenon is not clearly understood. This phenomenon happens three times in the case of the group $C_2 \times C_2 \times C_2$, six times in the case of the group $C_2 \times C_2 \times C_2 \times C_2$, eleven times in the case of the group $C_2 \times C_2 \times C_2 \times C_2 \times C_2$, and sixteen times in the case of the group $C_2 \times C_2 \times C_2 \times C_2 \times C_2 \times C_2$. Since the higher order $C_2 \times C_2 \times ...$ groups contain the lower order $C_2 \times C_2 \times ...$ groups, there will be such non-linear elimination equations in all higher order groups of this form, but it is unknown whether or not such equations become comparatively less frequent in the higher $C_2 \times C_2 \times ...$ groups. From calculation, your author can attest that no non-linear potential scaling parameter elimination equations exist in the groups $C_3 \times C_3$ or $C_4 \times C_4$ or C_2, C_3, C_4, C_5.

When we use the positive solution, $\theta = +\dfrac{\eta\delta}{\alpha}$, of the above (7.16), we get eight separate commutative algebras. These algebras are of the form:

$$\exp\left(\begin{bmatrix} a & b & c & d \\[6pt] \alpha b & a & \delta d & \dfrac{\alpha}{\delta}c \\[6pt] \eta c & \dfrac{\eta\delta}{\alpha}d & a & \dfrac{\alpha}{\delta}b \\[6pt] \dfrac{\eta\delta^2}{\alpha}d & \dfrac{\eta\delta}{\alpha}c & \delta b & a \end{bmatrix}\right) \quad : \{\alpha,\delta,\eta\} \neq 0 \quad (7.17)$$

The forms of the eight separate algebras arise from the eight permutations of setting the scaling parameters $\{\alpha, \delta, \eta\} = \pm 1$. Of these eight algebras, two algebras, the A_1 algebras, have three symmetric imaginary variables (three square roots of $+1$) and six algebras, the A_2 algebras, have one symmetric imaginary variable with two anti-symmetric imaginary variables (one square root of $+1$ and two square roots of -1). These algebras seem to play no part in the physical world; they keep themselves to themselves. The two A_1 algebras are algebraically isomorphic to each other, but, it seems (see later), reality cares not for algebraic isomorphism, and so we will count them as two separate algebras. The six A_2 algebras are algebraically isomorphic to each other. The isomorphic algebras differ from each other only by being written in different bases.

Non-commutative algebras in a commutative group:
When we take the negative solution of the above equation, (7.15), $\theta = -\dfrac{\eta \delta}{\alpha}$, we get eight non-commutative algebraic forms arising from the eight permutations of setting the scaling parameters $\{\alpha, \delta, \eta\} = \pm 1$. Of these eight algebras, two algebras have three anti-symmetric imaginary variables (three square roots of -1) and are quaternion type algebras that are associated with electromagnetism and six algebras have one anti-symmetric imaginary variable and two symmetric imaginary variables (one square root of -1 and two square roots of $+1$) and are A_3 type algebras that seem collectively to form the 4-dimensional space-time in which we sit. All from just one little minus sign. The six A_3 algebras are algebraically isomorphic, but we still count them as six separate algebras. The two quaternion algebras are algebraically isomorphic, but we count them as two separate algebras. These eight non-commutative algebras are of the form:

$$\exp\left(\begin{bmatrix} a & b & c & d \\ \alpha b & a & \delta d & \dfrac{\alpha}{\delta}c \\ \eta c & -\dfrac{\eta\delta}{\alpha}d & a & -\dfrac{\alpha}{\delta}b \\ -\dfrac{\eta\delta^2}{\alpha}d & \dfrac{\eta\delta}{\alpha}c & -\delta b & a \end{bmatrix}\right) : \{\alpha, \delta, \eta\} \neq 0 \quad (7.18)$$

The commutator of these algebras is:

$$AE - EA =$$

$$\begin{bmatrix} 0 & 2\dfrac{\delta\eta}{\alpha}(dg-ch) & 2\delta(bh-df) & 2\dfrac{\alpha}{\delta}(bg-cf) \\ 2\delta\eta(dg-ch) & 0 & 2\alpha(bg-cf) & 2\alpha(bh-df) \\ 2\delta\eta(bh-df) & -2\eta(bg-cf) & 0 & -2\eta(dg-ch) \\ -2\delta\eta(bg-cf) & 2\dfrac{\delta^2\eta}{\alpha}(bh-df) & -2\dfrac{\delta^2\eta}{\alpha}(dg-ch) & 0 \end{bmatrix} \quad (7.19)$$

We get the separate algebras by setting the three scaling parameters to the permutations of $\{-1, +1\}$. For example, if we set $\{\alpha = -1,\ \partial = -1,\ \eta = -1\}$, we get the quaternion algebra:

$$\mathbb{H} = \begin{bmatrix} a & b & c & d \\ -b & a & -d & c \\ -c & d & a & -b \\ -d & -c & b & a \end{bmatrix} \quad (7.20)$$

In this, we have the correspondence $b \to \hat{i},\ c \to j,\ d \to k$ with the usual quaternion notation:

$$\mathbb{H} = a + \hat{i}b + jc + kd \quad (7.21)$$

The other quaternion algebra, the anti-quaternions, $\mathbb{H}_{Anti}$, are given by $\{\alpha = -1,\ \partial = +1,\ \eta = -1\}$. We will later associate the anti-quaternions with anti-matter; we note that the difference between the two algebras is the sign of the δ scaling parameter; we will associate the scaling parameter δ, or its equivalent, with the charge of the electron.

In the quaternion case, the commutator becomes:

$$AE - EA =$$

$$\begin{bmatrix} 0 & -2(dg-ch) & -2(bh-df) & 2(bg-cf) \\ 2(dg-ch) & 0 & -2(bg-cf) & -2(bh-df) \\ 2(bh-df) & 2(bg-cf) & 0 & 2(dg-ch) \\ -2(bg-cf) & 2(bh-df) & -2(dg-ch) & 0 \end{bmatrix} \quad (7.22)$$

Which is, of course, a quaternion. This commutator is another way of writing $SU(2)$.

The non-commutative $C_2 \times C_2$ algebras:

Our concern is with the non-commutative algebras that derive from the finite groups. Taking the scaling parameters to be the permutations of $\{\alpha, \partial, \eta = \{+1, -1\}\}$ leads to eight separate algebras (only two algebraically non-isomorphic types).

The A_3 algebras are[39]:

---

[39] The names reflect, in alphabetical order, the symmetric/anti-symmetric variables.

$$SSA = \exp\left(\begin{bmatrix} a & b & c & d \\ b & a & d & c \\ c & -d & a & -b \\ -d & c & -b & a \end{bmatrix}\right)$$

$$SSA_{Anti} = \exp\left(\begin{bmatrix} a & b & c & d \\ b & a & -d & -c \\ c & d & a & b \\ -d & -c & b & a \end{bmatrix}\right)$$

(7.23)

$$SAS = \exp\left(\begin{bmatrix} a & b & c & d \\ b & a & d & c \\ -c & d & a & -b \\ d & -c & -b & a \end{bmatrix}\right)$$

$$SAS_{Anti} = \exp\left(\begin{bmatrix} a & b & c & d \\ b & a & -d & -c \\ -c & -d & a & b \\ d & c & b & a \end{bmatrix}\right)$$

(7.24)

$$ASS = \exp\left(\begin{bmatrix} a & b & c & d \\ -b & a & -d & c \\ c & -d & a & -b \\ d & c & b & a \end{bmatrix}\right)$$

$$ASS_{Anti} = \exp\left(\begin{bmatrix} a & b & c & d \\ -b & a & d & -c \\ c & d & a & b \\ d & -c & -b & a \end{bmatrix}\right)$$

(7.25)

The quaternion algebras are:

$$\mathbb{H} = \begin{bmatrix} a & b & c & d \\ -b & a & -d & c \\ -c & d & a & -b \\ -d & -c & b & a \end{bmatrix} \qquad \mathbb{H}_{Anti} = \begin{bmatrix} a & b & c & d \\ -b & a & d & -c \\ -c & -d & a & b \\ -d & c & -b & a \end{bmatrix}$$

$$(7.26)$$

Equivalent distributions of scaling parameters:

Above, by our choice of which scaling parameters to eliminate, we were driven to:

$$\begin{bmatrix} a & b & c & d \\ \alpha b & a & \delta d & \dfrac{\alpha}{\delta}c \\ \eta c & -\dfrac{\eta\delta}{\alpha}d & a & -\dfrac{\alpha}{\delta}b \\ -\dfrac{\eta\delta^2}{\alpha}d & \dfrac{\eta\delta}{\alpha}c & -\delta b & a \end{bmatrix} \qquad (7.27)$$

If we had chosen to eliminate other scaling parameters, we would have been driven to a different but equivalent matrix. The equivalence of the matrices is expressed by the equivalence of scaling parameters. Let us put $\varepsilon = \dfrac{\alpha}{\delta}$ into (7.27). We get:

$$\begin{bmatrix} a & b & c & d \\ \alpha b & a & \dfrac{\alpha}{\varepsilon}d & \varepsilon c \\ \eta c & -\dfrac{\eta}{\varepsilon}d & a & -\varepsilon b \\ -\dfrac{\alpha\eta}{\varepsilon^2}d & \dfrac{\eta}{\varepsilon}c & -\dfrac{\alpha}{\varepsilon}b & a \end{bmatrix} \qquad (7.28)$$

Setting the scaling parameters in (7.28) to the permutations of $\{-1,+1\}$ again leads to the eight above algebras but not the same algebra for the same permutation. Above, (7.20), setting $\{\alpha = -1, \partial = -1, \eta = -1\}$, led to the quaternion algebra; if we put the same permutation of scaling parameters into (7.28), we get:

$$\mathbb{H}_{Anti} = \begin{bmatrix} a & b & c & d \\ -b & a & d & -c \\ -c & -d & a & b \\ -d & c & -b & a \end{bmatrix} \qquad (7.29)$$

Setting $\alpha = -1, \delta = -1 \Rightarrow \varepsilon = \dfrac{\alpha}{\delta} = +1$ gives the quaternions, $\mathbb{H}$,

and so, with this choice of scaling parameter distribution, it is the ε scaling parameter which we will associate with charge of the electron – see (8.18).

Determinants of the A_3 algebras:

The norm of the algebras is given by the square root of the determinant. The norm is the distance function of the space. Those distance functions are:

$$
\begin{array}{lll}
SSA: & & dist^2 = a^2 - b^2 - c^2 + d^2 \\
SSA_{Anti}: & & dist^2 = a^2 - b^2 - c^2 + d^2 \\
SAS: & & dist^2 = a^2 - b^2 + c^2 - d^2 \\
SAS_{Anti}: & & dist^2 = a^2 - b^2 + c^2 - d^2 \\
ASS: & & dist^2 = a^2 + b^2 - c^2 - d^2 \\
ASS_{Anti}: & & dist^2 = a^2 + b^2 - c^2 - d^2 \\
\mathbb{H}: & & dist^2 = a^2 + b^2 + c^2 + d^2 \\
\mathbb{H}_{Anti}: & & dist^2 = a^2 + b^2 + c^2 + d^2
\end{array}
\qquad (7.30)
$$

Note on the Clifford algebra $Cl_{2,0}$:

The Clifford algebra, $Cl_{2,0}$, has the basis $\left\{1, \vec{e_1}, \vec{e_2}, \overrightarrow{e_{12}}\right\}$. $\overrightarrow{e_{12}}$ is a bi-vector. We have:

$$\vec{e_1}^2 = +1, \qquad \vec{e_2}^2 = +1$$
$$\overrightarrow{e_{12}}^2 = -1, \qquad \overrightarrow{e_{12}} = -\overrightarrow{e_{21}} \tag{7.31}$$

The four basis elements have the same algebraic properties as the variables in the A_3 algebras. The two vectors are square roots of plus unity which correspond to two symmetric matrix variables, and the bi-vector is a square root of minus unity which corresponds to an anti-symmetric matrix variable. Taking the SSA algebra as an example, we see that the matrix product of the two symmetric variables $\{b,c\}$ is an anti-symmetric variable $\{d\}$ and that the sign of the anti-symmetric variable is changed with the order of multiplication. The Clifford product of two 2-dimensional vectors is:

$$\vec{a}\vec{b} = a_1b_1 + a_2b_2 + (a_1b_2 - a_2b_1)\overrightarrow{e_{12}} \tag{7.32}$$

This is the sum of a scalar and a bi-vector. Within the SSA algebra, we have:

$$
\begin{bmatrix} 0 & a_1 & a_2 & 0 \\ a_1 & 0 & 0 & a_2 \\ a_2 & 0 & 0 & -a_1 \\ 0 & a_2 & -a_1 & 0 \end{bmatrix}
\begin{bmatrix} 0 & b_1 & b_2 & 0 \\ b_1 & 0 & 0 & b_2 \\ b_2 & 0 & 0 & -b_1 \\ 0 & b_2 & -b_1 & 0 \end{bmatrix}
$$
$$
= \begin{bmatrix} a_1b_1 + a_2b_2 & 0 & 0 & a_1b_2 - a_2b_1 \\ 0 & \sim & \sim & 0 \\ 0 & \sim & \sim & 0 \\ -(a_1b_2 - a_2b_1) & 0 & 0 & a_1b_1 + a_2b_2 \end{bmatrix} \tag{7.33}
$$

The inner product of the two vectors is the real variable, and the wedge product of the two vectors is the anti-symmetric imaginary variable. The Clifford product of two vectors is then traditionally written as:

$$\vec{a}\vec{b} = \vec{a}\cdot\vec{b} + \vec{a}\wedge\vec{b}$$
$$\vec{b}\vec{a} = \vec{a}\cdot\vec{b} - \vec{a}\wedge\vec{b}$$

(7.34)

In Clifford algebra, two vectors are parallel when they commute. We will see something very much like this later when we use non-commutativity to measure deviation from a geodesic.

The Clifford Algebra $Cl_{3,0}$ and spinors:

The Clifford algebra $Cl_{3,0}$ has the basis of one scalar, three vectors, three bi-vectors and one tri-vector:

$$\left\{\begin{array}{c} 1 \\ \vec{e_1}, \vec{e_2}, \vec{e_3} \\ \hline \vec{e_1}\vec{e_2}, \vec{e_1}\vec{e_3}, \vec{e_2}\vec{e_3} \\ \vec{e_1}\vec{e_2}\vec{e_3} \end{array}\right\}$$

(7.35)

The Clifford product within $Cl_{3,0}$ is:

$$\left(a\vec{e_1} + b\vec{e_2} + c\vec{e_3}\right)\left(x\vec{e_1} + y\vec{e_2} + z\vec{e_3}\right)$$
$$= (ax + by + cz) + (ay - bx)\vec{e_1}\vec{e_2}$$
$$+ (az - cx)\vec{e_1}\vec{e_3} + (bz - cy)\vec{e_2}\vec{e_3}$$
$$\equiv \vec{A}\cdot\vec{X} + \vec{A}\times\vec{X}$$

(7.36)

Each of the vectors $\{\vec{e_1}, \vec{e_2}, \vec{e_3}\}$ is a square root of plus unity, and all of the bi-vectors are non-commutative with each other $\{\vec{e_1}\vec{e_2} = -\vec{e_2}\vec{e_1}, \vec{e_1}\vec{e_3} = -\vec{e_3}\vec{e_1}, \vec{e_2}\vec{e_3} = -\vec{e_3}\vec{e_2}\}$ and therefore the bi-vectors are square roots of minus unity. The scalar, 1, is obviously a square root of plus unity, and the tri-vector is a square root of minus unity. We therefore associate this algebra with an 8-dimensional algebra with three symmetric imaginary variables and four anti-symmetric imaginary variables. Such algebras are within the group $C_2 \times C_2 \times C_2$. It is of interest that, "*..spinor spaces or spinor representations of the rotation group $SO(3)$ can be constructed within the Clifford algebra $Cl_{3,0}$.*"[40]

Clifford algebras in general:

In general, Clifford algebra deals with objects (vectors, bi-vectors, tri-vectors …) that are the square roots of plus unity or minus unity. We find such objects within the $C_2 \times C_2 \times$... groups as symmetric and anti-symmetric imaginary variables. We opine that the concept of bi-vectors and such within Clifford algebra are obfuscating distractions and that the symmetric and anti-symmetric matrices of the $C_2 \times C_2 \times$... groups are a mathematically tidier and more simple way of doing Clifford algebra. The $C_2 \times C_2 \times$... groups contain much more than is available within conventional Clifford algebra.[41]

---

[40] Pertti Lounesto, Clifford Algebras and Spinors, pg 53. ISBN: 0-521-00551-5
[41] See Dennis Morris The Naked Spinor – a rewrite of Clifford Algebra : ISBN: 978-1507817995

Chapter 8

Scaling Parameters

We have frequently referred to scaling parameters. We have told the reader from where they come, but we have written little of them. We now rectify that situation. Let us consider the 2-dimensional algebras. We have shown above, (3.31), that the determinant of an algebraic matrix form is the distance function of the geometric space that is associated with that algebra. In the case of the hyperbolic complex numbers, we have:

$$S_\lambda = \begin{bmatrix} a & b \\ \lambda b & a \end{bmatrix} \quad : \quad \lambda > 0$$

(8.1)

$$\det\left(\begin{bmatrix} a & b \\ \lambda b & a \end{bmatrix} \right) = a^2 - \lambda b^2$$

We normally take the 2-dimensional distance function of space-time to be:

$$d^2 = c^2 t^2 - z^2$$
$$\equiv t^2 - \frac{1}{c^2} z^2$$

(8.2)

We see that we have the scaling constant of space-time to be related to the physical constant that is the limiting velocity of the universe. This is really nothing new.

$$c_{LV} \equiv \frac{1}{\sqrt{\lambda}}$$

(8.3)

We have attached the subscript LV to the constant c to differentiate it from the velocity of light. As will become clearer later, it seems that the velocity of light is an electromagnetic constant and the limiting velocity of the universe is a space-time constant. We know that these two constants are observed, within space-time, to be equal. However, it seems that electromagnetism is a quaternion phenomenon and space-time is an A_3 algebra phenomenon. That space-time and electromagnetism come from different types of algebras is the basis of the assertion that the speed of light and the limiting velocity of the universe are not the same constant.

The physical constants of the universe:

It seems that the scaling parameters of various algebras are the physical constants of the universe. For decades, physicists have seen the speed of light as a scaling constant measuring the time axis against the space axis and so this idea is not new. This is an assertion that is backed by some evidence. However, your author admits that the evidence is not overwhelming. Although the evidence for this assertion is sketchy, we do at least now have an explanation for why physical constants exist and why there are the number of them that we observe. Your author knows of no other such proposal to explain the existence of the physical constants. Your author has no idea at all why the physical constants are of the particular values we observe but these obviously vary with the units we use.

Above, we have it that the scaling parameter of 2-dimensional space-time, $\mathbb{S}$, is the physical constant that is the limiting velocity of the universe, and so what about the other 2-dimensional division algebra, $\mathbb{C}$.

The momentum operator:

Let us look at the Euclidean complex numbers, $\mathbb{C}_\lambda$. We have:

$$\begin{bmatrix} 0 & 1 \\ -\lambda & 0 \end{bmatrix}^2 = \lambda \begin{bmatrix} -1 & 0 \\ 0 & -1 \end{bmatrix} \tag{8.4}$$

We will differentiate a complex function with respect to the imaginary axis.

$$\frac{\partial \begin{bmatrix} f(a,b) & g(a,b) \\ -\lambda g(a,b) & f(a,b) \end{bmatrix}}{\partial \begin{bmatrix} 0 & b \\ -\lambda b & 0 \end{bmatrix}} = \frac{1}{\begin{bmatrix} 0 & 1 \\ -\lambda & 0 \end{bmatrix}} \frac{\partial \begin{bmatrix} f(a,b) & g(a,b) \\ -\lambda g(a,b) & f(a,b) \end{bmatrix}}{\partial \begin{bmatrix} b & 0 \\ 0 & b \end{bmatrix}}$$

$$= \begin{bmatrix} 0 & -\dfrac{1}{\lambda} \\ 1 & 0 \end{bmatrix} \begin{bmatrix} \dfrac{\partial f}{\partial b} & \dfrac{\partial g}{\partial b} \\ -\lambda \dfrac{\partial g}{\partial b} & \dfrac{\partial f}{\partial b} \end{bmatrix} \qquad (8.5)$$

$$= \begin{bmatrix} 0 & -1 \\ \lambda & 0 \end{bmatrix} \begin{bmatrix} \dfrac{1}{\lambda} & 0 \\ 0 & \dfrac{1}{\lambda} \end{bmatrix} \begin{bmatrix} \dfrac{\partial f}{\partial b} & \dfrac{\partial g}{\partial b} \\ -\lambda \dfrac{\partial g}{\partial b} & \dfrac{\partial f}{\partial b} \end{bmatrix}$$

We see that the process of differentiation by an imaginary variable necessitates multiplication by the negative of the imaginary unit, $-i_\lambda$ and by the inverse of the scaling parameter, $\dfrac{1}{\lambda}$. We might, in other notation, write this differentiation operation as:

$$-i_\lambda \frac{1}{\lambda} \frac{\partial}{\partial b} \qquad (8.6)$$

Within quantum mechanics, we have the momentum operator:

$$p_x = -i\hbar \frac{\partial}{\partial x} \qquad (8.7)$$

Wherein $\hbar = \dfrac{h}{2\pi}$ is known as 'aitch-bar'. The constant h, known as Planck's constant, is a fundamental physical constant within quantum mechanics. We see that:

$$\hbar \equiv \frac{1}{\lambda} \qquad (8.8)$$

> *Aside:* We see a profound connection between the limiting velocity of the universe when $\lambda > 0$ and the quantum mechanical constant $\hbar$ when $\lambda < 0$.

Eigenfunctions and eigenvalues:

Well, it seems that the quantum mechanical momentum operator is no more than differentiate with respect to the imaginary variable, but, before we accept this, let us look further. Since the mathematical form of the 'differentiate with respect to the imaginary variable' operator is identical to the momentum operator, it comes as no surprise to find it has the same eigenfunctions and the same eigenvalues. The rotation matrix of the scaled complex numbers, $\mathbb{C}_\lambda$, is:

$$\exp\left(\begin{bmatrix} 0 & b \\ -\lambda b & 0 \end{bmatrix}\right) = \begin{bmatrix} \cos\left(\sqrt{\lambda}b\right) & \frac{1}{\sqrt{\lambda}}\sin\left(\sqrt{\lambda}b\right) \\ -\lambda\frac{1}{\sqrt{\lambda}}\sin\left(\sqrt{\lambda}b\right) & \cos\left(\sqrt{\lambda}b\right) \end{bmatrix} \tag{8.9}$$

Let us differentiate this rotation matrix of the $\mathbb{C}_\lambda$ algebra with respect to the imaginary variable. We will do it with an angle of $b = n\theta$:

$$\frac{\partial \begin{bmatrix} \cos\left(n\sqrt{\lambda}\theta\right) & \frac{1}{\sqrt{\lambda}}\sin\left(n\sqrt{\lambda}\theta\right) \\ -\lambda\frac{1}{\sqrt{\lambda}}\sin\left(n\sqrt{\lambda}\theta\right) & \cos\left(n\sqrt{\lambda}\theta\right) \end{bmatrix}}{\partial \begin{bmatrix} 0 & \theta \\ -\lambda\theta & 0 \end{bmatrix}} \tag{8.10}$$

$$= \begin{bmatrix} 0 & -\frac{1}{\lambda} \\ 1 & 0 \end{bmatrix} \begin{bmatrix} \frac{\partial}{\partial\theta}\cos\left(n\sqrt{\lambda}\theta\right) & \frac{1}{\sqrt{\lambda}}\frac{\partial}{\partial\theta}\sin\left(n\sqrt{\lambda}\theta\right) \\ -\lambda\frac{1}{\sqrt{\lambda}}\frac{\partial}{\partial\theta}\sin\left(n\sqrt{\lambda}\theta\right) & \frac{\partial}{\partial\theta}\cos\left(n\sqrt{\lambda}\theta\right) \end{bmatrix}$$

$$= \begin{bmatrix} 0 & -1 \\ \lambda & 0 \end{bmatrix} \begin{bmatrix} \frac{1}{\lambda} & 0 \\ 0 & \frac{1}{\lambda} \end{bmatrix} \begin{bmatrix} -n\sqrt{\lambda}\sin\left(n\sqrt{\lambda}\theta\right) & \frac{n\sqrt{\lambda}}{\sqrt{\lambda}}\cos\left(n\sqrt{\lambda}\theta\right) \\ -\lambda\frac{n\sqrt{\lambda}}{\sqrt{\lambda}}\cos\left(n\sqrt{\lambda}\theta\right) & -n\sqrt{\lambda}\sin\left(n\sqrt{\lambda}\theta\right) \end{bmatrix}$$

$$= \begin{bmatrix} n & 0 \\ 0 & n \end{bmatrix} \begin{bmatrix} \cos\left(n\sqrt{\lambda}\theta\right) & \frac{1}{\sqrt{\lambda}}\sin\left(n\sqrt{\lambda}\theta\right) \\ -\lambda\frac{1}{\sqrt{\lambda}}\sin\left(n\sqrt{\lambda}\theta\right) & \cos\left(n\sqrt{\lambda}\theta\right) \end{bmatrix}$$

(8.11)

We see that differentiation with respect to the imaginary variable of the rotation matrix with angle $n\theta$ simply multiplies the rotation matrix by n. We say that the rotation matrix with angle $n\theta$ is an eigenfunction of the 'differentiate with respect to the imaginary variable' operator. We say that the number, n, is the eigenvalue associated with that particular eigenfunction.

If, instead of $n\theta$, we use the angle $n\lambda\theta$, the rotation matrix will be multiplied by $n\lambda$, and, using other notation, we can write this as:

$$-i_\lambda \frac{1}{\lambda}\frac{\partial}{\partial\theta}\left(e^{i_\lambda n\lambda\theta}\right) = -i_\lambda \frac{1}{\lambda}i_\lambda n\lambda e^{i_\lambda n\lambda\theta} = ne^{i_\lambda n\lambda\theta} \qquad (8.12)$$

This is equivalent to the momentum operator:

$$-i\hbar\frac{\partial}{\partial x}\left(e^{i\frac{n}{\hbar}x}\right) = ne^{i\frac{n}{\hbar}x} \qquad (8.13)$$

It seems that we have the momentum operator and that the scaling parameter of the complex numbers, $\mathbb{C}_\lambda$, is the inverse of the physical constant $\hbar$. Of course, we are free to set the scaling constant to unity just as we often do with $\hbar$.

The 2-dimensional case for scaling parameters being physical constants is quite simple. The higher dimensional case is not so simple.

The nature of physical constants:

On Earth, we declare that the charge of an electron, the speed of light, and aitch-bar are physical constants. We mean that these are (some of) the basic constants of the universe. We say that the fine structure constant, α_{FS} :

$$\alpha_{FS} = \frac{e^2}{c\hbar} \qquad (8.14)$$

is a derived physical constant. On planet Zog, they say that the fine structure constant is one of the basic physical constants and that the charge of the electron is a derived physical constant:

$$e = \sqrt{\alpha_{FS} c \hbar} \qquad (8.15)$$

and so it does not surprise them to discover positrons. However, the Zoggians are in academic dispute with people on the planet Yog who assert that the electron charge divided by the electron's mass is a basic physical constant. What we all agree upon is the algebraic relation that connects the fine structure constant, the electron charge, the velocity of light and aitch-bar. The difference between the Earthling view and the Zoggian view is as the difference between a physical system described by polar co-ordinates and the same physical system described by Cartesian co-ordinates. In the same way that a co-ordinate system is arbitrary, so the choice of a particular set of basic physical constants is arbitrary. The algebraic relationship between those constants is invariant between the choices. Hence, we do not expect to have a particular set of basic constants in the grand unified theory of the universe but we do expect to have the algebraic relationship between those constants. This is exactly what we seem to get with the scaling parameters of a division algebra.

Quaternion scaling parameters:

We use the form of the quaternions derived above, (7.28):

$$\begin{bmatrix} a & b & c & d \\ \beta b & a & \dfrac{\beta}{\varepsilon}d & \varepsilon c \\ \eta c & -\dfrac{\eta}{\varepsilon}d & a & -\varepsilon b \\ -\dfrac{\beta\eta}{\varepsilon^2}d & \dfrac{\eta}{\varepsilon}c & -\dfrac{\beta}{\varepsilon}b & a \end{bmatrix} \tag{8.16}$$

This is a quaternion when $\{\beta<0,\ \varepsilon>0,\ \eta<0\}$. Setting three of the variables in turn to zero and squaring leads to[42]:

$$\beta\begin{bmatrix} b & 0 & 0 & 0 \\ 0 & b & 0 & 0 \\ 0 & 0 & b & 0 \\ 0 & 0 & 0 & b \end{bmatrix},\quad \eta\begin{bmatrix} c & 0 & 0 & 0 \\ 0 & c & 0 & 0 \\ 0 & 0 & c & 0 \\ 0 & 0 & 0 & c \end{bmatrix},\quad -\dfrac{\beta\eta}{\varepsilon^2}\begin{bmatrix} d & 0 & 0 & 0 \\ 0 & d & 0 & 0 \\ 0 & 0 & d & 0 \\ 0 & 0 & 0 & d \end{bmatrix} \tag{8.17}$$

Equating, by guesswork and with reference to the sentence below, also see (7.29), these scaling parameters with the physical constants:

$$\beta = c,\quad \varepsilon = e,\quad \eta = \alpha_{FS} \tag{8.18}$$

Gives:

$$c\begin{bmatrix} b & 0 & 0 & 0 \\ 0 & b & 0 & 0 \\ 0 & 0 & b & 0 \\ 0 & 0 & 0 & b \end{bmatrix},\quad \alpha_{FS}\begin{bmatrix} c & 0 & 0 & 0 \\ 0 & c & 0 & 0 \\ 0 & 0 & c & 0 \\ 0 & 0 & 0 & c \end{bmatrix},\quad -h\begin{bmatrix} d & 0 & 0 & 0 \\ 0 & d & 0 & 0 \\ 0 & 0 & d & 0 \\ 0 & 0 & 0 & d \end{bmatrix} \tag{8.19}$$

It would be lovely, at this point, to produce some electromagnetic formulae from the quaternions and show that the scaling parameters arise in exactly the same places as the physical constants with which we have identified them. At our present state of knowledge, we cannot do this, and so the assertion that the physical constants are scaling parameters is not yet satisfactorily shown.

---

[42] The existence of the multiplicative identity is what underlies this calculation.

What is what with scaling parameters:

Above, (8.17), we have seen that the imaginary d-variable of the quaternion scales by $-\dfrac{\beta\eta}{\varepsilon^2}$. So, it seems that the geometric scaling parameters are $\left\{\beta, \eta, -\dfrac{\beta\eta}{\varepsilon^2}\right\}$ rather than $\{\beta, \eta, \varepsilon\}$. Of course, varying ε will scale the imaginary d-variable, and so we are justified in picking which-ever set of scaling parameters we like.

Scaling parameters in higher dimensional spaces:

We postulate, with only one not very good reason to do so, that the 8-dimensional non-commutative $C_2 \times C_2 \times C_2$ algebras are connected to the strong force and that the 16 dimensional non-commutative $C_2 \times C_2 \times C_2 \times C_2$ algebras are connected to super-symmetric forces. If this is the case, there will be seven scaling parameters associated with seven physical constants for each algebraically distinct 8-dimensional algebra and fifteen scaling parameters associated with fifteen physical constants for each algebraically distinct 16-dimensional algebra, but we are now becoming extremely speculative.

In the 8-dimensional algebras, three of the scaling parameters are the same as the 4-dimensional scaling parameters because the 4-dimensional parameters are in the $C_2 \times C_2$ part of $C_2 \times C_2 \times C_2$. If we set the signs of the 4-dimensional scaling parameters to give an 8-dimensional 4-dimensional quaternion sub-algebra, then, because electromagnetism is a quaternion phenomenon, we will have electromagnetism within the 8-dimensional algebra. The scaling parameters of the seven imaginary variables are then:

$$b \sim P_{2,1}, \quad c \sim P_{3,1}, \quad d \sim \dfrac{P_{2,1} P_{3,1}}{\left(P_{2,4}\right)^2} \tag{8.20}$$

$$e \sim P_{5,1}, \quad f \sim \frac{P_{2,1}P_{5,1}}{\left(P_{2,6}\right)^2}, \quad g \sim \frac{P_{3,1}P_{5,1}}{\left(P_{3,7}\right)^2}, \quad h \sim \frac{P_{2,1}P_{3,1}P_{5,1}}{\left(P_{2,8}\right)^2\left(P_{3,7}\right)^2} \quad (8.21)$$

Above, (8.17) & (8.18), we associated the squared denominator, $\left(P_{2,4}\right)^2$ with electric charge (the charge of the electron). It would seem that in 8-dimensions, we have three more such charges $\left(P_{2,6}\right)^2, \left(P_{3,7}\right)^2, \left(P_{2,8}\right)^2$. This is the one not very good reason referred to above for postulating that the 8-dimensional non-commutative $C_2 \times C_2 \times C_2$ algebras are connected to the strong force. We think these might be the colour charges.

We opine that the 8-dimensional algebras are associated with a higher energy level than the 4-dimensional algebras; similarly, we opine that the 16-dimensional algebras are associated with a higher energy level than the 8-dimensional algebras.

Philosophical implications:
If we can get the physical constants from no more than division algebras (geometric spaces), then perhaps the universe is no more than these different empty spaces (numbers). Since numbers have to exist because, if they did not exist there would be zero of them, and zero is a number - it is one number; oh! there's another number – then the universe has to exist. The group C_2 has to exist because it is no more than the multiplicative relations between the numbers $\{-1,+1\}$, and all the $C_2 \times C_2 \times ...$ groups have to exist because they are the multiplicative relations between sets of $\{-1,+1\}$; for example:

$$C_2 \times C_2 \equiv \left\{ \begin{bmatrix} +1 \\ +1 \end{bmatrix}, \begin{bmatrix} +1 \\ -1 \end{bmatrix}, \begin{bmatrix} -1 \\ +1 \end{bmatrix}, \begin{bmatrix} -1 \\ -1 \end{bmatrix} \right\} \quad (8.22)$$

We are still a long way from being able to derive the universe from nothing more than the existence of numbers, but, although we have not yet shown it, we have large parts of electromagnetism, and, it seems, we have the 4-dimensional space-time in which we sit. Since we can derive these aspects of the universe, perhaps the rest is also in there.

Chapter 9

4-dimensional Commutation Relations and Lie Algebras

We have eight non-commutative algebras from $C_2 \times C_2$. We deal first with the two quaternion algebras. We have above shown that the commutator of the quaternion algebra

$$
\mathbb{H} = \begin{bmatrix} a & b & c & d \\ -b & a & -d & c \\ -c & d & a & -b \\ -d & -c & b & a \end{bmatrix}
\tag{9.1}
$$

is:

$$
AE - EA =
$$

$$
\begin{bmatrix}
0 & -2(dg-ch) & -2(bh-df) & 2(bg-cf) \\
2(dg-ch) & 0 & -2(bg-cf) & -2(bh-df) \\
2(bh-df) & 2(bg-cf) & 0 & 2(dg-ch) \\
-2(bg-cf) & 2(bh-df) & -2(dg-ch) & 0
\end{bmatrix}
\tag{9.2}
$$

This is a composition of individual imaginary variable matrices. For example:

$$
\begin{bmatrix} 0 & 1 & 0 & 0 \\ -1 & 0 & 0 & 0 \\ 0 & 0 & 0 & -1 \\ 0 & 0 & 1 & 0 \end{bmatrix}
\begin{bmatrix} 0 & 0 & 1 & 0 \\ 0 & 0 & 0 & 1 \\ -1 & 0 & 0 & 0 \\ 0 & -1 & 0 & 0 \end{bmatrix}
=
\begin{bmatrix} 0 & 0 & 0 & 1 \\ 0 & 0 & -1 & 0 \\ 0 & 1 & 0 & 0 \\ -1 & 0 & 0 & 0 \end{bmatrix}
\tag{9.3}
$$

$$\begin{bmatrix} 0 & 0 & 1 & 0 \\ 0 & 0 & 0 & 1 \\ -1 & 0 & 0 & 0 \\ 0 & -1 & 0 & 0 \end{bmatrix} \begin{bmatrix} 0 & 1 & 0 & 0 \\ -1 & 0 & 0 & 0 \\ 0 & 0 & 0 & -1 \\ 0 & 0 & 1 & 0 \end{bmatrix} = \begin{bmatrix} 0 & 0 & 0 & -1 \\ 0 & 0 & 1 & 0 \\ 0 & -1 & 0 & 0 \\ 1 & 0 & 0 & 0 \end{bmatrix} \qquad (9.4)$$

This is conventionally written as:

$$\hat{i}\hat{j} = k \qquad : \qquad \hat{j}\hat{i} = -k$$
$$\left[\hat{i}, j\right] = 2k \qquad\qquad (9.5)$$

The other well-known quaternion commutation relations are similarly calculated. The quaternion commutation relations are exactly the same as the $SU(2)$ commutation relations, but there is a conceptual difference in that the quaternions are a division algebra which we call quaternion space, $\mathbb{H}$, and $SU(2)$ is seen as being a rotation within $\mathbb{C}^2$ space.

The two forms of quaternion each have three anti-symmetric imaginary variables. Indeed, they are the only two possible ways of fitting together three 4-dimensional anti-symmetric variables to form a division algebra. We repeat the quaternion commutation relations for completeness.

$$\mathbb{H}: \qquad \left[\hat{i}, j\right] = k \qquad : \qquad \left[j, \hat{i}\right] = -k$$
$$\left[j, k\right] = \hat{i} \qquad : \qquad \left[k, j\right] = -\hat{i} \qquad\qquad (9.6)$$
$$\left[k, \hat{i}\right] = j \qquad : \qquad \left[\hat{i}, k\right] = -j$$

The anti-quaternions have:

$$\mathbb{H}_{Anti}: \qquad \left[\hat{i}, j\right] = -k \qquad : \qquad \left[j, \hat{i}\right] = k$$
$$\left[j, k\right] = -\hat{i} \qquad : \qquad \left[k, j\right] = \hat{i} \qquad\qquad (9.7)$$
$$\left[k, \hat{i}\right] = -j \qquad : \qquad \left[\hat{i}, k\right] = j$$

These two algebras are algebraically isomorphic. They are isomorphic as Lie algebras. We can think of the two algebras as being a clockwise and an anti-clockwise version of the one algebra. However, although they are the same algebra written in two different bases, they are two algebras. They both exist; they both came from the mathematics. We think (see later) this is connected to anti-matter and we take the view that we have here both $SU(2)$ and the anti-matter version of $SU(2)_{Anti}$. We have not yet shown it, but there is something much more than this in these two algebras; within these two algebras lies the electromagnetic tensor; if only one of these algebras existed, we would not be able to derive the conventional electromagnetic tensor from them.

The commutator of two space-time boosts:

The Lorentz group, $SO(3,1)$, has the remarkable property that the commutator of two space-time boosts is a spatial rotation. The A_3 algebras have the same property, but we need to explain this in more detail. The A_3 algebras each have two symmetric imaginary variables and one anti-symmetric imaginary variable. The exponential of an anti-symmetric imaginary variable matrix is a spatial rotation matrix associated with Euclidean trigonometric functions. For example:

$$\exp\left(\begin{bmatrix} 0 & b & 0 & 0 \\ -b & 0 & 0 & 0 \\ 0 & 0 & 0 & -b \\ 0 & 0 & b & 0 \end{bmatrix}\right) =$$

$$\begin{bmatrix} \cos\sqrt{b^2} & \sin\sqrt{b^2} & 0 & 0 \\ -\sin\sqrt{b^2} & \cos\sqrt{b^2} & 0 & 0 \\ 0 & 0 & \cos\sqrt{b^2} & -\sin\sqrt{b^2} \\ 0 & 0 & \sin\sqrt{b^2} & \cos\sqrt{b^2} \end{bmatrix} \qquad (9.8)$$

The exponential of a symmetric imaginary variable matrix is a space-time rotation matrix – a boost; it is a rotation associated with hyperbolic trigonometric functions. For example:

$$\exp\left(\begin{bmatrix} 0 & b & 0 & 0 \\ b & 0 & 0 & 0 \\ 0 & 0 & 0 & -b \\ 0 & 0 & -b & 0 \end{bmatrix}\right) =$$

$$\begin{bmatrix} \cosh\sqrt{b^2} & \sinh\sqrt{b^2} & 0 & 0 \\ \sinh\sqrt{b^2} & \cosh\sqrt{b^2} & 0 & 0 \\ 0 & 0 & \cosh\sqrt{b^2} & -\sinh\sqrt{b^2} \\ 0 & 0 & -\sinh\sqrt{b^2} & \cosh\sqrt{b^2} \end{bmatrix}$$

(9.9)

We thus associate anti-symmetric imaginary variable matrices with spatial rotations and symmetric imaginary variables with space-time rotations (boosts). Within each of the A_3 algebras, we have it that the commutator of two symmetric matrices is an anti-symmetric matrix. For example:

$$\begin{bmatrix} 0 & 1 & 0 & 0 \\ 1 & 0 & 0 & 0 \\ 0 & 0 & 0 & -1 \\ 0 & 0 & -1 & 0 \end{bmatrix}\begin{bmatrix} 0 & 0 & 1 & 0 \\ 0 & 0 & 0 & 1 \\ 1 & 0 & 0 & 0 \\ 0 & 1 & 0 & 0 \end{bmatrix} = \begin{bmatrix} 0 & 0 & 0 & 1 \\ 0 & 0 & 1 & 0 \\ 0 & -1 & 0 & 0 \\ -1 & 0 & 0 & 0 \end{bmatrix}$$

$$\begin{bmatrix} 0 & 0 & 1 & 0 \\ 0 & 0 & 0 & 1 \\ 1 & 0 & 0 & 0 \\ 0 & 1 & 0 & 0 \end{bmatrix}\begin{bmatrix} 0 & 1 & 0 & 0 \\ 1 & 0 & 0 & 0 \\ 0 & 0 & 0 & -1 \\ 0 & 0 & -1 & 0 \end{bmatrix} = \begin{bmatrix} 0 & 0 & 0 & -1 \\ 0 & 0 & -1 & 0 \\ 0 & 1 & 0 & 0 \\ 1 & 0 & 0 & 0 \end{bmatrix}$$

(9.10)

Thus, the A_3 algebras each possess this 'the commutator of two boosts is a spatial rotation' property that we normally associate with the Lorentz group. We point out that these spatial and space-time rotations are 4-dimensional 2-dimensional (sorry again) rotations and are different from the 2-dimensional rotations with which we are familiar.

The A_3 commutation relations:

The six A_3 algebras are algebraically isomorphic. The commutation relations are isomorphic in the Lie algebra sense. The commutation relations come as a matrix, one for each algebra, as is the case with the quaternions, but they are more clearly seen in the conventional notation. Those commutation relations are:

$$SSA: \quad \left[i_S, j_S\right] = k_A \quad : \quad \left[j_S, i_S\right] = -k_A$$

$$\left[i_S, k_A\right] = j_S \quad : \quad \left[k_A, i_S\right] = -j_S \qquad (9.11)$$

$$\left[j_S, k_A\right] = -i_S \quad : \quad \left[k_A, j_S\right] = i_S$$

In matrix form:

$$SSA_{COMM} =$$

$$\begin{bmatrix} 0 & 2(dg-ch) & 2(bh-df) & 2(bg-cf) \\ 2(dg-ch) & 0 & 2(bg-cf) & 2(bh-df) \\ 2(bh-df) & -2(bg-cf) & 0 & -2(dg-ch) \\ -2(bg-cf) & 2(bh-df) & -2(dg-ch) & 0 \end{bmatrix} \quad (9.12)$$

$$SSA_{Anti}: \quad \left[i_S, j_S\right] = -k_A \quad : \quad \left[j_S, i_S\right] = k_A$$

$$\left[i_S, k_A\right] = -j_S \quad : \quad \left[k_A, i_S\right] = j_S \qquad (9.13)$$

$$\left[j_S, k_A\right] = i_S \quad : \quad \left[k_A, j_S\right] = -i_S$$

$$SAS: \quad \left[i_S, j_A \right] = k_S \quad : \quad \left[j_A, i_S \right] = -k_S$$

$$\left[i_S, k_S \right] = j_A \quad : \quad \left[k_S, i_S \right] = -j_A \qquad (9.14)$$

$$\left[j_A, k_S \right] = i_S \quad : \quad \left[k_S, j_A \right] = -i_S$$

$$SAS_{Anti}: \quad \left[i_S, j_A \right] = -k_S \quad : \quad \left[j_A, i_S \right] = k_S$$

$$\left[i_S, k_S \right] = -j_A \quad : \quad \left[k_S, i_S \right] = j_A \qquad (9.15)$$

$$\left[j_A, k_S \right] = -i_S \quad : \quad \left[k_S, j_A \right] = i_S$$

$$ASS: \quad \left[i_A, j_S \right] = k_S \quad : \quad \left[j_S, i_A \right] = -k_S$$

$$\left[i_A, k_S \right] = -j_S \quad : \quad \left[k_S, i_A \right] = j_S \qquad (9.16)$$

$$\left[j_S, k_S \right] = -i_A \quad : \quad \left[k_S, j_S \right] = i_A$$

$$ASS_{Anti}: \quad \left[i_A, j_S \right] = -k_S \quad : \quad \left[j_S, i_A \right] = k_S$$

$$\left[i_A, k_S \right] = j_S \quad : \quad \left[k_S, i_A \right] = -j_S \qquad (9.17)$$

$$\left[j_S, k_S \right] = i_A \quad : \quad \left[k_S, j_S \right] = -i_A$$

The reader will note that we have three pairs of algebras comprised of an algebra and its anti-algebra. The reader will see that we have labelled the $\left\{ \hat{i}, j, k \right\}$ with subscripts to indicate that the variable is symmetric or anti-symmetric.

Because a single A_3 algebra has only one anti-symmetric imaginary variable, we do not have the commutator of two anti-symmetric imaginary variables anywhere; furthermore, within a single A_3 algebra, we have only the commutator of the two symmetric imaginary variables; we do not have the set of commutators of three symmetric imaginary variables.

The Lorentz group:

The Lorentz group, $SO(3,1)$, is represented by six matrices. Of these six matrices, three are symmetric matrices:

$$K_1 = \begin{bmatrix} 0 & 1 & 0 & 0 \\ 1 & 0 & 0 & 0 \\ 0 & 0 & 0 & 0 \\ 0 & 0 & 0 & 0 \end{bmatrix}, \ K_2 = \begin{bmatrix} 0 & 0 & 1 & 0 \\ 0 & 0 & 0 & 0 \\ 1 & 0 & 0 & 0 \\ 0 & 0 & 0 & 0 \end{bmatrix}, \ K_3 = \begin{bmatrix} 0 & 0 & 0 & 1 \\ 0 & 0 & 0 & 0 \\ 0 & 0 & 0 & 0 \\ 1 & 0 & 0 & 0 \end{bmatrix} \quad (9.18)$$

and three are anti-symmetric matrices:

$$J_1 = \begin{bmatrix} 0 & 0 & 0 & 0 \\ 0 & 0 & 0 & 0 \\ 0 & 0 & 0 & -i \\ 0 & 0 & i & 0 \end{bmatrix}, \ J_2 = \begin{bmatrix} 0 & 0 & 0 & 0 \\ 0 & 0 & 0 & i \\ 0 & 0 & 0 & 0 \\ 0 & -i & 0 & 0 \end{bmatrix}, \ J_3 = \begin{bmatrix} 0 & 0 & 0 & 0 \\ 0 & 0 & -i & 0 \\ 0 & i & 0 & 0 \\ 0 & 0 & 0 & 0 \end{bmatrix} \quad (9.19)$$

The three K_i matrices are seen as being associated with three 2-dimensional space-time rotations (boosts) in 4-dimensional space-time. Each K_i matrix represents a space-time rotation in a 2-dimensional space-time 'plane'. The three different K_i matrices represent space-time rotations in three different 2-dimensional space-time 'planes'. The three J_i matrices are associated with three 2-dimensional spatial rotations in what is conventionally seen as the 3-dimensional sub-space of 4-dimensional space-time.

Lorentz group $\rightarrow C_2 \times C_2$

We associate the K_i matrices with symmetric imaginary variable matrices from the A_3 algebras. We associate the J_i matrices with anti-symmetric imaginary variable matrices from the A_3 algebras:

$$K_1 \rightarrow \begin{bmatrix} 0 & 1 & 0 & 0 \\ 1 & 0 & 0 & 0 \\ 0 & 0 & 0 & 1 \\ 0 & 0 & 1 & 0 \end{bmatrix} \rightarrow \begin{bmatrix} 0 & 1 & 0 & 0 \\ 1 & 0 & 0 & 0 \\ 0 & 0 & 0 & -1 \\ 0 & 0 & -1 & 0 \end{bmatrix} \quad (9.20)$$

$$K_2 \rightarrow \begin{bmatrix} 0 & 0 & 1 & 0 \\ 0 & 0 & 0 & 1 \\ 1 & 0 & 0 & 0 \\ 0 & 1 & 0 & 0 \end{bmatrix} \rightarrow \begin{bmatrix} 0 & 0 & 1 & 0 \\ 0 & 0 & 0 & -1 \\ 1 & 0 & 0 & 0 \\ 0 & -1 & 0 & 0 \end{bmatrix} \quad (9.21)$$

$$K_3 \rightarrow \begin{bmatrix} 0 & 0 & 0 & 1 \\ 0 & 0 & 1 & 0 \\ 0 & 1 & 0 & 0 \\ 1 & 0 & 0 & 0 \end{bmatrix} \rightarrow \begin{bmatrix} 0 & 0 & 0 & 1 \\ 0 & 0 & -1 & 0 \\ 0 & -1 & 0 & 0 \\ 1 & 0 & 0 & 0 \end{bmatrix} \quad (9.22)$$

Note that both of the pair of matrices are the same matrix in a different basis.

$$J_1 \rightarrow \begin{bmatrix} 0 & 1 & 0 & 0 \\ -1 & 0 & 0 & 0 \\ 0 & 0 & 0 & -1 \\ 0 & 0 & 1 & 0 \end{bmatrix} \rightarrow \begin{bmatrix} 0 & 1 & 0 & 0 \\ -1 & 0 & 0 & 0 \\ 0 & 0 & 0 & 1 \\ 0 & 0 & -1 & 0 \end{bmatrix} \quad (9.23)$$

$$J_2 \rightarrow \begin{bmatrix} 0 & 0 & 1 & 0 \\ 0 & 0 & 0 & 1 \\ -1 & 0 & 0 & 0 \\ 0 & -1 & 0 & 0 \end{bmatrix} \rightarrow \begin{bmatrix} 0 & 0 & 1 & 0 \\ 0 & 0 & 0 & -1 \\ -1 & 0 & 0 & 0 \\ 0 & 1 & 0 & 0 \end{bmatrix} \quad (9.24)$$

$$J_3 \rightarrow \begin{bmatrix} 0 & 0 & 0 & 1 \\ 0 & 0 & 1 & 0 \\ 0 & -1 & 0 & 0 \\ -1 & 0 & 0 & 0 \end{bmatrix} \rightarrow \begin{bmatrix} 0 & 0 & 0 & 1 \\ 0 & 0 & -1 & 0 \\ 0 & 1 & 0 & 0 \\ -1 & 0 & 0 & 0 \end{bmatrix} \quad (9.25)$$

Again, both of the pair of matrices are the same matrix in a different basis.

The K_i matrices alone:

The commutation relations of the K_i matrices with each other are:

$$[K_1, K_2] = iJ_3$$
$$[K_1, K_3] = -iJ_2 \qquad (9.26)$$
$$[K_2, K_3] = iJ_1$$

There are two symmetric matrices in each A_3 algebra just as there are two K_i in each commutator bracket. We associate each commutator with a different one of the A_3 algebras. This means that the K_i matrices are not fixed but change from algebra to algebra. This is no more than a change of basis under a unitary transformation (the eigenvalues are the same for both matrices). What is fixed about the A_3 matrices associated with the K_i matrices is the fact that they are symmetric.

What we have in essence within these commutation relations of the Lorentz group is that the commutator of two space-time rotations in different 2-dimensional space-time planes is not a space-time rotation but a spatial rotation. This is, in essence, what we have in the A_3 algebras, as shown above, (9.10).

The J_i matrices alone:

The J_i matrices have the commutation relations of $SU(2)$ with each other. These occur in the quaternion algebra. We identify the J_i matrices with the anti-symmetric imaginary variables within the non-commutative $C_2 \times C_2$ algebras. Since there are six such matrices, again, we have that the Lorentz group matrices are each associated with two different $C_2 \times C_2$ matrices, but, again, the only difference between the two matrices is the basis in which they are written. Again, the bases are connected by a unitary transformation.

The $\{J_i, K_i\}$ together:

We will deal with the commutation relations between the J_i matrices and the K_i matrices in two parts. Firstly, the commutation relations between the J_n matrices and the K_n matrices are:

$$[J_1, K_1] = [J_2, K_2] = [J_2, K_2] = 0 \qquad (9.27)$$

Within any particular $C_2 \times C_2$ algebra, we do not have an imaginary variable that is both symmetric and anti-symmetric at the same time, obviously, and so commutation relations between such matrices do not exist.

Secondly, the remaining commutation relations of the Lorentz group are:

$$[J_1, K_2] = iK_3 \qquad [J_1, K_3] = -iK_2$$
$$[J_2, K_1] = -iK_3 \qquad [J_2, K_3] = iK_1 \qquad (9.28)$$
$$[J_3, K_1] = iK_2 \qquad [J_3, K_2] = -iK_1$$

These correspond to the commutators of an anti-symmetric matrix with a symmetric matrix within a particular A_3 algebra. In each case, we have that the commutator of an anti-symmetric matrix with a symmetric matrix is a symmetric matrix. This matches the A_3 algebras, but it is not universally true that the commutator of an anti-symmetric matrix with a symmetric matrix is a symmetric matrix. There are individual symmetric matrices within the matrices of the A_3 algebras that commute with individual anti-symmetric matrices within the matrices of the A_3 algebras, but such individual matrices never occur together in the same algebra.

Aside: We take the view that the $\hat{i}$s in the matrices of the conventionally presented Lorentz group are not an essential part of the Lorentz group but are merely a notational device. We take the view that the commutation relations are the essence of the Lorentz group. We take the same view of $SU(2)$.

$SU(2)$ & $SO(3,1)$ and the $C_2 \times C_2$ algebras:

The Classical Lie algebras are not the same as the division algebra Lie algebras. Firstly, there is a conceptual difference regarding the nature of the space that underlies the commutation relations that are the classical Lie algebras and the nature of space that underlies the division algebras. Secondly, although the complete set of commutation relations of $SU(2)$ are found within each single quaternion algebra, the complete set of commutation relations of $SO(3,1)$ are not found within a single A_3 algebra but are scattered among the six A_3 algebras.

From quantum physics to macroscopic physics:

We see that we get $SO(3,1)$ only by aggregating together the six A_3 algebras. Such aggregation is not an algebraic operation, and throughout the world it would have pedantic mathematicians throwing themselves from the roofs of mathematics buildings rather than countenance it. However, as will become clearer later, it seems that 'macroscopic physics' as done by macroscopic observers, rather than 'microscopic quantum physics, as done by microscopic observers, is done by such aggregation. It seems that each division algebra is quantum physics (with commutation relations) and that these are aggregated together to produce macroscopic physics. Such aggregation destroys the algebras and with them any meaningful concept of multiplication[43]. By this means, such aggregation destroys the commutation relations, and so it is that we move from the quantum world to the non-quantum macroscopic world. We will see that such aggregation produces the distance function of the space in which we sit, and we will see that such aggregation produces tensors. (Tensors do not exist within division algebras.) It seems that physics is based in two types of mathematics. The first type is the division algebras which correspond to the quantum world, and the second type corresponds to the macroscopic non-quantum world. That second type of mathematics arises from the division algebras by aggregation[44]. The Lorentz group, $SO(3,1)$, is the first example of such macroscopic aggregation that we have met. The Lorentz group, $SO(3,1)$, does not exist in its totality in any 4-dimensional division algebra although there is a double cover of it in the 8-dimensional $C_2 \times C_2 \times C_2$ algebras; it arises through the aggregation of the 4-dimensional A_3 algebras. The nature of that aggregation seems to be taking all the bits from each algebra to form a coherent whole.

The Lie group $SO(3,1)$ is usually thought of as a quantum mechanical phenomenon; we do not see it in macroscopic space; within the space-time in which we sit, we do not see the commutator of two boosts to be a spatial rotation. This is simply because we do not see 4-dimensional rotation in our space-time.

---

[43] Proper multiplication has to be closed; a duck mated with a duck must produce a duck and not a frog.
[44] So far, we do not know the rules controlling this aggregation.

Pauli Matrices and the A_3 Algebras

The 2×2 Pauli matrices are:

$$\sigma_x = \begin{bmatrix} 0 & 1 \\ 1 & 0 \end{bmatrix}, \quad \sigma_y = \begin{bmatrix} 0 & -i \\ i & 0 \end{bmatrix}, \quad \sigma_z = \begin{bmatrix} 1 & 0 \\ 0 & -1 \end{bmatrix} \quad (10.1)$$

The 2×2 Pauli matrices are hermitian matrices with the eigenvalues $\{+1, -1\}$. The eigenvalues correspond to the up spin or down spin of the electron. The 2×2 Pauli matrices are all square roots of plus unity. The eigenvectors of the 2-dimensional Pauli matrices are:

$$Eigenvectors(\sigma_x) = \left\{ \begin{bmatrix} 1 \\ 1 \end{bmatrix}, \begin{bmatrix} -1 \\ 1 \end{bmatrix} \right\}$$

$$Eigenvectors(\sigma_y) = \left\{ \begin{bmatrix} 1 \\ i \end{bmatrix}, \begin{bmatrix} 1 \\ -i \end{bmatrix} \right\} \quad (10.2)$$

$$Eigenvectors(\sigma_z) = \left\{ \begin{bmatrix} 1 \\ 0 \end{bmatrix}, \begin{bmatrix} 0 \\ 1 \end{bmatrix} \right\}$$

The Pauli matrices are the same matrix written in three different bases. They are connected by unitary transformations. We have:

$$U_{x \to z}^\dagger \sigma_x U_{x \to z} = \frac{1}{\sqrt{2}} \begin{bmatrix} 1 & 1 \\ -1 & 1 \end{bmatrix} \begin{bmatrix} 0 & 1 \\ 1 & 0 \end{bmatrix} \frac{1}{\sqrt{2}} \begin{bmatrix} 1 & -1 \\ 1 & 1 \end{bmatrix} = \begin{bmatrix} 1 & 0 \\ 0 & -1 \end{bmatrix} \quad (10.3)$$

We have:

$$U_{y \to z}^\dagger \sigma_y U_{y \to z} = \frac{1}{\sqrt{2}} \begin{bmatrix} -i & 1 \\ i & 1 \end{bmatrix} \begin{bmatrix} 0 & -i \\ i & 0 \end{bmatrix} \frac{1}{\sqrt{2}} \begin{bmatrix} -i & i \\ 1 & 1 \end{bmatrix} = \begin{bmatrix} 1 & 0 \\ 0 & -1 \end{bmatrix} \quad (10.4)$$

We have the 2×2 Pauli matrix multiplicative relations:

$$\sigma_x\sigma_y = -\sigma_y\sigma_x = \pm i\sigma_z$$
$$\sigma_y\sigma_z = -\sigma_z\sigma_y = \pm i\sigma_x \qquad (10.5)$$
$$\sigma_z\sigma_x = -\sigma_x\sigma_z = \pm i\sigma_y$$

This means that the Pauli matrices are the basis vectors of the 8-dimensional Clifford algebra, $Cl_{3,0}$.

4-dimensional Pauli matrices:
The 2×2 Pauli matrices have elements that are complex. We can take the view that any 2×2 matrix with complex numbers, $\mathbb{C}$, as elements is really a 4×4 matrix with real numbers as elements; algebraically the 2×2 matrix is isomorphic to the 4×4 matrix. We can write the Pauli matrices as 4×4 matrices by using the block multiplication properties of matrices and the following equivalences:

$$1 \equiv \begin{bmatrix} 1 & 0 \\ 0 & 1 \end{bmatrix}, \quad -1 \equiv \begin{bmatrix} -1 & 0 \\ 0 & -1 \end{bmatrix}, \quad i \equiv \begin{bmatrix} 0 & 1 \\ -1 & 0 \end{bmatrix} \qquad (10.6)$$

We have:

$$_4\sigma_x = \begin{bmatrix} 0 & 0 & 1 & 0 \\ 0 & 0 & 0 & 1 \\ 1 & 0 & 0 & 0 \\ 0 & 1 & 0 & 0 \end{bmatrix}, \quad _4\sigma_y = \begin{bmatrix} 0 & 0 & 0 & -1 \\ 0 & 0 & 1 & 0 \\ 0 & 1 & 0 & 0 \\ -1 & 0 & 0 & 0 \end{bmatrix} \qquad (10.7)$$

$$_4\sigma_z = \begin{bmatrix} 1 & 0 & 0 & 0 \\ 0 & 1 & 0 & 0 \\ 0 & 0 & -1 & 0 \\ 0 & 0 & 0 & -1 \end{bmatrix} \qquad (10.8)$$

Although this works, it gives only part of the story. It is better to simply view the 4×4 form of the Pauli matrices as the symmetric matrices from the A_3 algebras. There are six such matrices. We will get the three Pauli matrices and three anti-Pauli matrices that we think might be associated with anti-matter. We have no interest in the symmetric matrices with elements on only the leading diagonal.

4×4 symmetric matrices:

The symmetric matrices of the A_3 algebras are:

$$BM = \begin{bmatrix} 0 & 1 & 0 & 0 \\ 1 & 0 & 0 & 0 \\ 0 & 0 & 0 & -1 \\ 0 & 0 & -1 & 0 \end{bmatrix} \quad \& \quad B = \begin{bmatrix} 0 & 1 & 0 & 0 \\ 1 & 0 & 0 & 0 \\ 0 & 0 & 0 & 1 \\ 0 & 0 & 1 & 0 \end{bmatrix} \qquad (10.9)$$

$$CM = \begin{bmatrix} 0 & 0 & 1 & 0 \\ 0 & 0 & 0 & -1 \\ 1 & 0 & 0 & 0 \\ 0 & -1 & 0 & 0 \end{bmatrix} \quad \& \quad C = \begin{bmatrix} 0 & 0 & 1 & 0 \\ 0 & 0 & 0 & 1 \\ 1 & 0 & 0 & 0 \\ 0 & 1 & 0 & 0 \end{bmatrix} \qquad (10.10)$$

$$DM = \begin{bmatrix} 0 & 0 & 0 & 1 \\ 0 & 0 & -1 & 0 \\ 0 & -1 & 0 & 0 \\ 1 & 0 & 0 & 0 \end{bmatrix} \quad \& \quad D = \begin{bmatrix} 0 & 0 & 0 & 1 \\ 0 & 0 & 1 & 0 \\ 0 & 1 & 0 & 0 \\ 1 & 0 & 0 & 0 \end{bmatrix} \qquad (10.11)$$

We claim that these are Pauli matrices. The 4×4 symmetric matrices have eigenvalues $\{+1, +1, -1, -1\}$. The 4×4 symmetric matrices are all square roots of plus unity. The eigenvectors of the 4×4 symmetric matrices are:

$$Eigenvectors\,(B) = \left\{ \begin{bmatrix} 1 \\ -1 \\ 0 \\ 0 \end{bmatrix}, \begin{bmatrix} 0 \\ 0 \\ 1 \\ -1 \end{bmatrix}, \begin{bmatrix} 0 \\ 0 \\ 1 \\ 1 \end{bmatrix}, \begin{bmatrix} 1 \\ 1 \\ 0 \\ 0 \end{bmatrix} \right\} \qquad (10.12)$$

$$\text{Eigenvectors}\,(C) = \left\{ \begin{bmatrix} 0 \\ 1 \\ 0 \\ 1 \end{bmatrix}, \begin{bmatrix} 1 \\ 0 \\ 1 \\ 0 \end{bmatrix}, \begin{bmatrix} 0 \\ -1 \\ 0 \\ 1 \end{bmatrix}, \begin{bmatrix} 1 \\ 0 \\ -1 \\ 0 \end{bmatrix} \right\}$$

$$\text{Eigenvectors}\,(D) = \left\{ \begin{bmatrix} -1 \\ 0 \\ 0 \\ 1 \end{bmatrix}, \begin{bmatrix} 0 \\ -1 \\ 1 \\ 0 \end{bmatrix}, \begin{bmatrix} 1 \\ 0 \\ 0 \\ 1 \end{bmatrix}, \begin{bmatrix} 0 \\ 1 \\ 1 \\ 0 \end{bmatrix} \right\}$$

(10.13)

$$\text{Eigenvectors}\,(BM) = \left\{ \begin{bmatrix} 0 \\ 0 \\ 1 \\ 1 \end{bmatrix}, \begin{bmatrix} 1 \\ -1 \\ 0 \\ 0 \end{bmatrix}, \begin{bmatrix} 1 \\ 1 \\ 0 \\ 0 \end{bmatrix}, \begin{bmatrix} 0 \\ 0 \\ -1 \\ 1 \end{bmatrix} \right\}$$

$$\text{Eigenvectors}\,(CM) = \left\{ \begin{bmatrix} -1 \\ 0 \\ 1 \\ 0 \end{bmatrix}, \begin{bmatrix} 0 \\ 1 \\ 0 \\ 1 \end{bmatrix}, \begin{bmatrix} 0 \\ -1 \\ 0 \\ 1 \end{bmatrix}, \begin{bmatrix} 1 \\ 0 \\ 1 \\ 0 \end{bmatrix} \right\}$$

(10.14)

$$\text{Eigenvectors}\,(DM) = \left\{ \begin{bmatrix} 0 \\ -1 \\ 1 \\ 0 \end{bmatrix}, \begin{bmatrix} 1 \\ 0 \\ 0 \\ 1 \end{bmatrix}, \begin{bmatrix} 0 \\ 1 \\ 1 \\ 0 \end{bmatrix}, \begin{bmatrix} -1 \\ 0 \\ 0 \\ 1 \end{bmatrix} \right\}$$

The 4×4 symmetric matrices are the same matrix written in different bases. They are connected to each other by unitary transformations.

Multiplicative relations of the 4×4 symmetric matrices:
Any two 2×2 Pauli matrices multiply together to make the third Pauli matrix multiplied by $i = \sqrt{-1} = \begin{bmatrix} 0 & 1 \\ -1 & 0 \end{bmatrix}$. The 4×4 versions of $\sqrt{-1}$ are the six anti-symmetric imaginary matrices of the $C_2 \times C_2$ algebras. Those six anti-symmetric matrices are:

$$AB_1 = \begin{bmatrix} 0 & 1 & 0 & 0 \\ -1 & 0 & 0 & 0 \\ 0 & 0 & 0 & -1 \\ 0 & 0 & 1 & 0 \end{bmatrix}, \quad AB_2 = \begin{bmatrix} 0 & 1 & 0 & 0 \\ -1 & 0 & 0 & 0 \\ 0 & 0 & 0 & 1 \\ 0 & 0 & -1 & 0 \end{bmatrix} \quad (10.15)$$

$$AC_1 = \begin{bmatrix} 0 & 0 & 1 & 0 \\ 0 & 0 & 0 & -1 \\ -1 & 0 & 0 & 0 \\ 0 & 1 & 0 & 0 \end{bmatrix}, \quad AC_2 = \begin{bmatrix} 0 & 0 & 1 & 0 \\ 0 & 0 & 0 & 1 \\ -1 & 0 & 0 & 0 \\ 0 & -1 & 0 & 0 \end{bmatrix} \quad (10.16)$$

$$AD_1 = \begin{bmatrix} 0 & 0 & 0 & 1 \\ 0 & 0 & -1 & 0 \\ 0 & 1 & 0 & 0 \\ -1 & 0 & 0 & 0 \end{bmatrix}, \quad AD_2 = \begin{bmatrix} 0 & 0 & 0 & 1 \\ 0 & 0 & 1 & 0 \\ 0 & -1 & 0 & 0 \\ -1 & 0 & 0 & 0 \end{bmatrix} \quad (10.17)$$

The multiplicative relations of the symmetric matrices above are:

$$\begin{aligned} DM.C &= -C.DM = AB_1 \\ CM.D &= -D.CM = AB_2 \\ BM.D &= -D.BM = AC_1 \\ DM.B &= -B.DM = AC_2 \\ CM.B &= -B.CM = AD_1 \\ BM.C &= -C.BM = AD_2 \end{aligned} \qquad (10.18)$$

We demonstrate:

$$\begin{bmatrix} 0 & 0 & 0 & 1 \\ 0 & 0 & -1 & 0 \\ 0 & -1 & 0 & 0 \\ 1 & 0 & 0 & 0 \end{bmatrix} \begin{bmatrix} 0 & 0 & 1 & 0 \\ 0 & 0 & 0 & 1 \\ 1 & 0 & 0 & 0 \\ 0 & 1 & 0 & 0 \end{bmatrix} = \begin{bmatrix} 0 & 1 & 0 & 0 \\ -1 & 0 & 0 & 0 \\ 0 & 0 & 0 & -1 \\ 0 & 0 & 1 & 0 \end{bmatrix} \qquad (10.19)$$

$$= \begin{bmatrix} 0 & 1 & 0 & 0 \\ -1 & 0 & 0 & 0 \\ 0 & 0 & 0 & 1 \\ 0 & 0 & -1 & 0 \end{bmatrix} \begin{bmatrix} 1 & 0 & 0 & 0 \\ 0 & 1 & 0 & 0 \\ 0 & 0 & -1 & 0 \\ 0 & 0 & 0 & -1 \end{bmatrix} \qquad (10.20)$$

This is of the form:

$$\sigma_x^4 \sigma_y^4 = i\sigma_z^4 \qquad (10.21)$$

Why can we not simultaneously know $\{S_x, S_y, S_z\}$:

The A_3 algebras above correspond to six geometric spaces. It seems that we do not sit in a single space but that we sit in a superposition of six spaces. These six spaces come in three pairs comprised of an algebra and its anti-algebra. The algebras are isomorphic; they are the same algebra written in different bases.

When we measure the z-component of the spin of an electron, we say that the superposition collapses into an eigenvector of $\frac{\hbar}{2}\sigma_z$ and that we cannot know the x-component of spin or the y-component of spin simultaneously with the z-component of spin. We know that the eigenvectors of σ_z are the same as the eigenvectors of $\{\sigma_x, \sigma_y\}$ except that they are written in a different basis. It is most strange that we can know only one of the three components of spin at a time, but we now see why. Each 4×4 Pauli matrix is associated one of the three A_3 algebras. These three algebras are in different bases, and so the three Pauli matrices are the same matrix in different bases. Thus we now understand, at least as far as spin is concerned, the perplexing quantum phenomena that is mutually exclusive measurement. We see that, when it comes to spin, wave function collapse is no more than picking out the base (algebra) in which to measure. This picking might be done by way the observer sets up her observational apparatus.

Another way to view this phenomenon of mutually exclusive measurement is to accept that we are trying to observe a 4-dimensional rotation (spin) in a space-time that allows only 2-dimensional rotations. We are unable to know all three angle variables simultaneously because to do so would be to observe 4-dimensional rotation. The best we can do is to know one of the three angle variables at a time.

Limitations of the understanding:

We would have liked to have derived the whole of quantum field theory from the $C_2 \times C_2 \times \ldots$ algebras. Unfortunately, at this stage of our knowledge, we do not know how to do that.

Chapter 11

The Distance Function of our Space-time

Aggregating algebras together:

In this chapter, we are going to aggregate together aspects of the A_3 algebras. Such aggregation is not mathematics as we know mathematics. We do not understand the rules of algebra aggregation, if there are any, and so we do not really know what we are doing. Mathematicians cannot claim that what we do is incorrect, but they can claim that it is not mathematics. It might well be physics.

We think that aggregation is the equivalent of taking expectation values. Physicists take expectation values every day in quantum physics. We think we are doing the same except that we take expectation spaces, expectation distance functions, expectation tensors etc. Istead of only expectation values, we take expectation forms of other mathematical objects.

The justification for using aggregation is that it works; the results correspond to the observed macroscopic physical world. We think that the mathematics of division algebras corresponds to the observed microscopic quantum world. When we aggregate algebras together, we move from division algebras to an aggregate of those division algebras. We think that aggregating together algebras is moving from the quantum world to the macroscopic world. It might be that the rules of algebra aggregation depend upon the nature of the observations of the macroscopic world, but we do not know. We are embarrassed to so aggregate algebras together; in mathematical terms, it simply isn't cricket, but it does seem to win the match.

Alternatively, we can say that we are taking the average of the division algebras. In quantum mechanics, the average is called the expectation value. In this view, aggregation of division algebras is no more than calculating the expectation value.

Introduction to the distance function:

Physicists will tell us that the observed distance function of the space-time in which we sit is:

$$d^2 = c^2t^2 - x^2 - y^2 - z^2 \qquad (11.1)$$

Wherein c is no more than a factor that depends upon the units we use to measure space and time (metres and seconds). We seek to calculate the distance function:

$$d^2 = t^2 - x^2 - y^2 - z^2 \qquad (11.2)$$

wherein we have set $c = 1$.

There are eight non-commutative algebras that derive from the $C_2 \times C_2$ group. Two of these are quaternions which are, as we will see later, associated with electromagnetism. The other six seem to be concerned with the space-time in which we sit. We have seen the Pauli matrices are within these algebras, and we have seen the Lorentz group, $SO(3,1)$, is within the aggregate of these six algebras.

If the space-time in which we sit is really (at least[45]) six spaces, how do they relate together? Well, we are outside of 'proper' mathematics now. Normally, a mathematician would work within a single algebra. An aggregation of six algebraic objects from different algebras is not an algebraic object. Of course, the six algebras are the same algebra written in six different bases, but that does not mean we are doing 'proper' mathematics if we sum them. We are stumbling around in unmapped places. None-the-less, if that is how nature works, we must seek to make some sense of these unmapped places.

With thought, we realise that do not need to assume that nature works in 'improper mathematical ways'. We need to assume only that a macroscopic 'clodhopping' observer observes in 'improper mathematical ways'. It might be that only a macroscopic observer

---

[45] Your author has the lurking suspicion that the 8-dimensional spaces might manifest themselves in 4-dimensional guise by pairing two dimensions together, but his understanding is unclear on this point.

aggregates different algebras and sees the results and that, microscopically, nature sticks to proper mathematics.

> *Aside:* Of course, if we have a vector of an algebra written in a particular basis, $\vec{A}$, and we have a vector of the same algebra written in a different basis, $\vec{B}$, we can add these two vectors by first writing one of the vectors in the same basis as the other vector. However, nature seems not to work this way. Perhaps more accurately, the way we make macroscopically observations seems not to work this way.

The distance function of our space-time:

A simple view of how the six A_3 spaces 'fit together' is that the distance through these six spaces is simply the sum of the six individual distances through each space. Looking back to (7.30) leads to:

$$SUM \begin{cases} dist^2 = t^2 - x^2 - y^2 + z^2 \\ dist^2 = t^2 - x^2 - y^2 + z^2 \\ dist^2 = t^2 - x^2 + y^2 - z^2 \\ dist^2 = t^2 - x^2 + y^2 - z^2 \\ dist^2 = t^2 + x^2 - y^2 - z^2 \\ dist^2 = t^2 + x^2 - y^2 - z^2 \end{cases} = 2(3t^2 - x^2 - y^2 - z^2) \quad (11.3)$$

The 2 is just a doubling up because we have counted both the distance through space and through anti-space, or it is just a factor which we can forget. The 3 gets absorbed in the different units which we use to measure space and time. There we have it – simple! The distance through 4-dimensional space-time is just the sum of the distances through the six A_3 spaces. We think of this as an expectation distance function. Simple it might be, but it is very profound for it seems to be part of the difference between quantum physics and classical physics.

We might say that the distance function of our 4-dimensional space-time is the expectation distance function of the A_3 division algebras.

We point out that the two quaternion algebras (which we think are electromagnetism) and the six A_3 spaces are the only 4-dimensional non-commutative spaces. If we could not derive the distance function from them, all would be lost. That the only non-commutative and non-electromagnetic algebras add to exactly what we observe is remarkable. We have not picked out what we needed from a large selection of possible algebras; we have not omitted any algebras to suit our aims; we have taken exactly what the mathematics gave us, and it gives the required distance function. These algebras, as pointed out earlier, also hold the Lorentz group $SO(3,1)$. Everything fits.

Quantum space and quantum gravity:
We derived both the distance function and the Lorentz group from the A_3 algebras by aggregation. We associate such aggregation with macroscopic physics. We associate each division algebra with microscopic quantum physics. It seems that the individual A_3 algebras might be quantum space, and thus might be associated with quantum gravity, and that such space has the distance function with signature $(+,+,-,-)$ in contradistinction to the macroscopic distance function signature $(+,-,-,-)$.

Curvature and the metric?
The conceptual idea underlying general relativity is that gravity corresponds to the metric tensor varying from point to point in 4-dimensional space-time. Within an individual algebra, the distance function cannot vary from point to point without destroying the algebra, and so we cannot have quantum gravity of the same form as macroscopic gravity. The sum of six algebras is not an algebra, and so there is nothing to destroy if the metric tensor varies from point to point. Above, we have derived the distance function of the space-time in which we sit on the assumption that the six spaces we added are

all scaled at the same rate. There is no reason known to your author to assume this is necessarily the case[46]. If each space and its' anti-space are scaled equally but the three pairs are scaled differently, we get the sum:

$$
SUM \begin{cases}
dist^2 = t^2 - x^2 - y^2 + z^2 \\
dist^2 = t^2 - x^2 - y^2 + z^2 \\
dist^2 = \phi\left(t^2 - x^2 + y^2 - z^2\right) \\
dist^2 = \phi\left(t^2 - x^2 + y^2 - z^2\right) \\
dist^2 = \kappa\left(t^2 + x^2 - y^2 - z^2\right) \\
dist^2 = \kappa\left(t^2 + x^2 - y^2 - z^2\right)
\end{cases} \tag{11.4}
$$

$$
= 2\begin{pmatrix} (1+\phi+\kappa)t^2 - (1+\phi-\kappa)x^2 \\ -(1-\phi+\kappa)y^2 - (-1+\phi+\kappa)z^2 \end{pmatrix}
$$

Of course, we could scale each of the six spaces differently and claim that the resulting distance function is the canonical form of the general relativity metric. At any point, P, of a differential manifold, M, there exists a co-ordinate system in which the metric takes its canonical form and the first derivatives all vanish[47].

At our present level of understanding, we have no idea whether or not gravity or general relativity has anything to do with the above scaling factors, but, since the sum of algebras is not an algebra, there is no algebraic reason why the relative scaling factors, and thus the distance function, cannot vary from point to point within that sum of algebras. We will look at gravity in more detail later in this book. We think general relativity gravity has nothing to do with the above scaling factors; we include them because they might be important in

---

[46] We opine that the two quaternion spaces are scaled at the same rate because one of the scaling parameters, ε, is the charge of the electron, and so the 'electromagnetic distance scale' is set for both algebras – Hm! does this make sense?

[47] Carroll, Space-time and geometry, pg 73.

some way that we do not yet understand. They might be dark matter; who knows?

Aside: Summing the 2 algebraically isomorphic commutative 3-dimensional distance functions gives a 1-dimensional total distance function:

$$SUM \begin{Bmatrix} a^3 - b^3 + c^3 + 3abc \\ a^3 + b^3 - c^3 + 3abc \end{Bmatrix} = 2a^3 + 6abc \qquad (11.5)$$

We note that no 3-dimensional distance function can 'fit inside' our 4-dimensional space-time distance function in the way that the two 2-dimensional division algebra distance functions can 'fit inside' it.

Unbalanced physics from balanced mathematics:

The distribution of minus signs and plus signs is unbalanced in the macroscopic distance function of the space-time in which we sit; there are three minus signs and only one plus sign. This corresponds to an imbalance in the number of spatial dimensions and the number of time dimensions. Remarkably, this imbalance is derived from six balanced distance functions. The imbalance could not be avoided; there are only so many symmetric and anti-symmetric 4×4 matrices that can be used to form an algebra. We began with the $C_2 \times C_2$ group in which everything seemed to be in balance, and we were led to this imbalance. The imbalance is intrinsic to the mathematics.

2-dimensional rotations:

We have seen above (circa (3.22)) that a 2-dimensional rotation that does not affect the third and fourth dimension is a 2-dimensional rotation in a 2-dimensional space that is a separate space in its own right and not a 2-dimensional sub-space of a 4-dimensional space. We observe six such separate 2-dimensional spaces within the space-time in which we sit. Three of these six 2-dimensional spaces are space-time spaces and three are space-space spaces.

Let us take a pair of A_3 algebras (an algebra and its anti-algebra). Let us set all the variables to zero except one and then add the two algebras. We get (see (7.23)):

$$SSA + SSA_{Anti} = \begin{bmatrix} 0 & b & 0 & 0 \\ b & 0 & 0 & 0 \\ 0 & 0 & 0 & -b \\ 0 & 0 & -b & 0 \end{bmatrix} + \begin{bmatrix} 0 & b & 0 & 0 \\ b & 0 & 0 & 0 \\ 0 & 0 & 0 & b \\ 0 & 0 & b & 0 \end{bmatrix}$$

$$= \begin{bmatrix} 0 & 2b & 0 & 0 \\ 2b & 0 & 0 & 0 \\ 0 & 0 & 0 & 0 \\ 0 & 0 & 0 & 0 \end{bmatrix} \tag{11.6}$$

Forgetting the 2 and taking the exponential gives:

$$\exp \begin{bmatrix} 0 & b & 0 & 0 \\ b & 0 & 0 & 0 \\ 0 & 0 & 0 & 0 \\ 0 & 0 & 0 & 0 \end{bmatrix} = \begin{bmatrix} \cosh b & \sinh b & 0 & 0 \\ \sinh b & \cosh b & 0 & 0 \\ 0 & 0 & 0 & 0 \\ 0 & 0 & 0 & 0 \end{bmatrix} \tag{11.7}$$

We say that this is a 2-dimensional rotation (boost). We say the 4-dimensional spaces have thrown out a 2-dimensional space.

$$\begin{bmatrix} \cosh b & \sinh b \\ \sinh b & \cosh b \end{bmatrix} \tag{11.8}$$

Subtracting an anti-algebra from its corresponding algebra has a similar effect.

With another algebra, we get:

$$ASS + ASS_{Anti} = \begin{bmatrix} 0 & b & 0 & 0 \\ -b & 0 & 0 & 0 \\ 0 & 0 & 0 & -b \\ 0 & 0 & b & 0 \end{bmatrix} + \begin{bmatrix} 0 & b & 0 & 0 \\ -b & 0 & 0 & 0 \\ 0 & 0 & 0 & b \\ 0 & 0 & -b & 0 \end{bmatrix} \tag{11.9}$$

$$= \begin{bmatrix} 0 & 2b & 0 & 0 \\ -2b & 0 & 0 & 0 \\ 0 & 0 & 0 & 0 \\ 0 & 0 & 0 & 0 \end{bmatrix} \quad (11.10)$$

And:

$$\exp \begin{bmatrix} 0 & b & 0 & 0 \\ -b & 0 & 0 & 0 \\ 0 & 0 & 0 & 0 \\ 0 & 0 & 0 & 0 \end{bmatrix} = \begin{bmatrix} \cos b & \sin b & 0 & 0 \\ -\sin b & \cos b & 0 & 0 \\ 0 & 0 & 0 & 0 \\ 0 & 0 & 0 & 0 \end{bmatrix} \quad (11.11)$$

We say that this too is a 2-dimensional rotation. Similarly taking all the algebras and the different variables will give all six of the observed 2-dimensional rotations in all six planes.

A different view of the 2-dimensional rotations we observe in our 4-dimensional space-time is that, because only one of each of the 2-dimensional algebras emerges from the C_2 finite group, when the isomorphic algebras are aggregated together, the 2-dimensional algebras are not broken by the aggregation. Thus, they survive intact whereas all 4-dimensional algebras are broken. By 'luck'[48] the distance functions of the 2-dimensional algebras fit into the 4-dimensional distance function of space-time, and so we see 2-dimensional rotations within our 4-dimensional space-time but no 4-dimensional rotations. The 4-dimensional rotations having been broken by aggregation.

Pulling it together:
We have, with very simple and 'obvious' assumptions derived the distance function of the 4-dimensional space-time in which we sit from nothing more than the group $C_2 \times C_2$. There is nothing in conventional mathematics to say that we can add the distance

---

[48] It is nothing to do with luck. It is because C_2 is a sub-group of $C_2 \times C_2$.

functions of different algebras or of the same algebra in different bases; we just took a common sense view, but, other than this, we have not contrived the result. There is nothing in mathematics to say that we can add the variables of an algebra and its anti-algebra, but when we do it, we seem to get a fit with observation. Is the above mathematics really the physical universe? It is perhaps too early to say. Later in this book, we will derive general relativity from these algebras to go alongside the electromagnetism we are going to derive shortly from the quaternions and the special relativity that fell out of the group C_2. The derivation can be questioned, but it seems to work.

Quaternion space:

Adding the distance functions of the two quaternion algebras gives:

$$d_{quatternion}^2 = 2\left(a^2 + b^2 + c^2 + d^2\right) \qquad (11.12)$$

We will soon associate the quaternions with electromagnetism[49]. It would seem that electromagnetic space is different from the space-time in which we sit. Imagine an object moving in a straight line through electromagnetic space. It would appear to move in a curve when seen from space-time. Observers in space-time would say that a force was acting upon that object. Is this the nature of classical forces, or is force something to do with non-commutativity (see later), or is force something to do with the independent scaling of the six A_3 algebras?

The absence of quantitisation in classical physics:

Above, we have seen that each of the six A_3 spaces is quantitised with a select few of the $SO(3,1)$ commutation relations. If we add

---

[49] Physicists previous to ourselves have associated the quaternions with electromagnetism because they associate the Lie group $SU(2)$ with electromagnetism. Further, Maxwell used quaternions to write the first version of the Maxwell equations.

the six spaces in the way we have added the distance functions, we get a matrix that is not a division algebra. Multiplying elements of the matrix sum of the A_3 algebras together does not produce other elements of that matrix sum; therefore, there are no commutation relations associated with the matrix sum of the A_3 algebras. By taking the sum of the algebras, we have lost quantitisation. We will later argue that 'microscopic' physics happens in a single division algebra space and is hence quantitised but that 'macroscopic' physics is observed in a sum of division algebra spaces and is hence not quantitised.

Chapter 12

Quaternions and Non-commutative Differentiation

In an earlier chapter, we differentiated the complex numbers, $\mathbb{C}$, - see (5.5). The reader might recall that we took constant matrices out of the differential to the left in that chapter. Since $\mathbb{C}$ is a commutative algebra, it matters not whether we take constant matrices out of the differential to the left or constant matrices out of the differential to the right. However, when it comes to non-commutative algebras such as the quaternions or the A_3 algebras, to which side we take a constant matrix from the differential does matter. Within non-commutative algebras, we get two differentials, which we refer to as the left differential, d_L, and the right differential, d_R. The left differential is formed by taking all constant matrices from the differential to the left, and the right differential is formed by taking all constant matrices from the differential to the right[50].

Having formed the left and right differential, we form half the sum of the two differentials, which we call the E field, and we form half the difference of the two differentials, which we call the B field. We have the two matrices:

$$E_{\mu\nu} = \frac{1}{2}\left(d_L + d_R\right)$$
$$\overset{LR}{B}_{\mu\nu} = \frac{1}{2}\left(d_L - d_R\right)$$

(12.1)

If we were to reverse the order of the differentials in the B field, we would be defining the B field to act in the opposite direction with

---

[50] For this kind of differentiation, we must give some credit to Peter Michael Jack: Physical Space as a Quaternion Structure 1: Maxwell Equations – A Brief Note arXiv.math-ph.0307038v1 18 Jul 2003. Also, Jack cites C. J. Joly 1905 : A Manual of Quaternions : Art 57 pp 74-77. We should also mention David Hestenes who developed differentiation within Clifford algebras.

respect to the E field. The choice is arbitrary, and, if we wish, we are free to define the B field as:

$$\overset{RL}{B}_{\mu\nu} = \frac{1}{2}(d_R - d_L)$$

(12.2)

$$\overset{RL}{B} = -\overset{LR}{B}$$

We adopt the convention that, in forming the B field, we will always subtract the right differential from the left differential, and so the B field will always be defined as given in (12.1).

We will shortly deduce electromagnetism from the quaternions using non-commutative differentiation. Because we have defined the B field as (12.1), and because the curl of a vector field is defined in a particular direction, and because the magnetic field is defined as the curl of a vector field rather than as minus the curl of a vector field, we will get the conventional magnetic field as the B field of the quaternions. If we reversed only the definition of the B field, we would get the conventional magnetic field as the B field of the anti-quaternions rather than the quaternions; similarly, if we reversed only the definition of the curl or redefined the magnetic field to be minus the curl, we would get the conventional magnetic field from the anti-quaternions rather than from the quaternions. If we reversed only two of the definitions, we would again get the conventional magnetic field as the B field of the quaternions. Perhaps the reader might like to look at the B field of the anti-quaternions in (14.61). The E field of the quaternions is the same as the E field of the anti-quaternions, and so, the difference between the quaternions and the anti-quaternions is just the direction of the magnetic field[51]. By opting for the definition of the B field as (12.1), we have opted to use the quaternions rather than the anti-quaternions to express the conventional form of electromagnetism.

General comment on types of differentiation:
As the reader is doubtless aware, in life, we meet many types of puddings. We also meet many types of differentiation; there is no god

---

[51] We think the anti-quaternions are anti-matter, but see later.

given type of differentiation to which we must all pay homage. The proof of the differentiation is, as with puddings, in the results. So it must be with the non-commutative differentiation that we present here. We think this non-commutative differentiation does produce the proper results in the form of 4-dimensional curls, divergence, and gradient. It is for the reader to make their own assessment. We will add that this non-commutative differentiation concurs with Clifford algebra differentiation although it goes further that Clifford algebra differentiation and it is mathematically tidier than Clifford algebra differentiation.

How to differentiate in non-commutative algebras:
In this chapter, to demonstrate the technique to the reader, we will show the cumbersome details of the procedure of non-commutative differentiation. This means that we have 4×4 matrix equations filling pages. Even so, the actual differentials are so cumbersome that we cannot fit them on to the page and we will write them, not as matrices but as a list of elements of those matrices. Because we are working in the quaternion algebra, every matrix we calculate will be a quaternion, and, once we have the basic matrix form, we need give only the four elements of the top row of the quaternion to specify the whole matrix. If we want the whole matrix, we simply put these four top row elements into the appropriate positions in the 4×4 quaternion matrix, with, of course, the minus signs in the proper places. We remind the reader that the (equally scaled) quaternion matrix is:

$$\mathbb{H} = \begin{bmatrix} t & x & y & z \\ -x & t & -z & y \\ -y & z & t & -x \\ -z & -y & x & t \end{bmatrix} \tag{12.3}$$

Non-commutative differentiation:
Because this chapter is a pedagogical demonstration of a mathematical technique, we will ignore the scaling parameters; they

add complications which 'clog up' the page without giving any advantage. We will differentiate the quaternion potential:

$$\Phi_{Pot} = \begin{bmatrix} \phi & A_x & A_y & A_z \\ -A_x & \phi & -A_z & A_y \\ -A_y & A_z & \phi & -A_x \\ -A_z & -A_y & A_x & \phi \end{bmatrix} \; : \; \begin{matrix} \phi(t,x,y,z) \\ A_x(t,x,y,z) \\ A_y(t,x,y,z) \\ A_z(t,x,y,z) \end{matrix} \quad (12.4)$$

with respect to the quaternion:

$$Q_{(t,x,y,z)} = \begin{bmatrix} t & x & y & z \\ -x & t & -z & y \\ -y & z & t & -x \\ -z & -y & x & t \end{bmatrix} \quad (12.5)$$

To fit the mathematics on to the page, we will sometimes use the notation:

$$\frac{\partial \begin{bmatrix} \phi & A_x & A_y & A_z \\ -A_x & \phi & -A_z & A_y \\ -A_y & A_z & \phi & -A_x \\ -A_z & -A_y & A_x & \phi \end{bmatrix}}{\partial \begin{bmatrix} t & x & y & z \\ -x & t & -z & y \\ -y & z & t & -x \\ -z & -y & x & t \end{bmatrix}} = \frac{\partial \Phi_{Pot}}{\partial Q_{(t,x,y,z)}} \quad (12.6)$$

To either side, the differential is the sum of the individual variable differentials:

$$\frac{\partial \Phi_{Pot}}{\partial Q_{(t,x,y,z)}} = -\frac{\partial \begin{bmatrix} \phi & A_x & A_y & A_z \\ -A_x & \phi & -A_z & A_y \\ -A_y & A_z & \phi & -A_x \\ -A_z & -A_y & A_x & \phi \end{bmatrix}}{\partial \begin{bmatrix} t & 0 & 0 & 0 \\ 0 & t & 0 & 0 \\ 0 & 0 & t & 0 \\ 0 & 0 & 0 & t \end{bmatrix}} + \frac{\partial \begin{bmatrix} \phi & A_x & A_y & A_z \\ -A_x & \phi & -A_z & A_y \\ -A_y & A_z & \phi & -A_x \\ -A_z & -A_y & A_x & \phi \end{bmatrix}}{\partial \begin{bmatrix} 0 & x & 0 & 0 \\ -x & 0 & 0 & 0 \\ 0 & 0 & 0 & -x \\ 0 & 0 & x & 0 \end{bmatrix}}$$

(12.7)

$$+ \frac{\partial \begin{bmatrix} \phi & A_x & A_y & A_z \\ -A_x & \phi & -A_z & A_y \\ -A_y & A_z & \phi & -A_x \\ -A_z & -A_y & A_x & \phi \end{bmatrix}}{\partial \begin{bmatrix} 0 & 0 & y & 0 \\ 0 & 0 & 0 & y \\ -y & 0 & 0 & 0 \\ 0 & -y & 0 & 0 \end{bmatrix}} + \frac{\partial \begin{bmatrix} \phi & A_x & A_y & A_z \\ -A_x & \phi & -A_z & A_y \\ -A_y & A_z & \phi & -A_x \\ -A_z & -A_y & A_x & \phi \end{bmatrix}}{\partial \begin{bmatrix} 0 & 0 & 0 & z \\ 0 & 0 & -z & 0 \\ 0 & z & 0 & 0 \\ -z & 0 & 0 & 0 \end{bmatrix}}$$

(12.8)

Differentiation to the left:
We will do each piece of this differentiation one at a time. We will take the matrices from the denominators to the left. We have:

$$\frac{\partial \begin{bmatrix} \phi & A_x & A_y & A_z \\ -A_x & \phi & -A_z & A_y \\ -A_y & A_z & \phi & -A_x \\ -A_z & -A_y & A_x & \phi \end{bmatrix}}{\partial \begin{bmatrix} t & 0 & 0 & 0 \\ 0 & t & 0 & 0 \\ 0 & 0 & t & 0 \\ 0 & 0 & 0 & t \end{bmatrix}} = \begin{bmatrix} \frac{\partial \phi}{\partial t} & \frac{\partial A_x}{\partial t} & \frac{\partial A_y}{\partial t} & \frac{\partial A_z}{\partial t} \\ -\frac{\partial A_x}{\partial t} & \frac{\partial \phi}{\partial t} & -\frac{\partial A_z}{\partial t} & \frac{\partial A_y}{\partial t} \\ -\frac{\partial A_y}{\partial t} & \frac{\partial A_z}{\partial t} & \frac{\partial \phi}{\partial t} & -\frac{\partial A_x}{\partial t} \\ -\frac{\partial A_z}{\partial t} & -\frac{\partial A_y}{\partial t} & \frac{\partial A_x}{\partial t} & \frac{\partial \phi}{\partial t} \end{bmatrix}$$

(12.9)

121

That was simple. The next bit is slightly more complicated. We have:

$$\frac{\partial \begin{bmatrix} \phi & A_x & A_y & A_z \\ -A_x & \phi & -A_z & A_y \\ -A_y & A_z & \phi & -A_x \\ -A_z & -A_y & A_x & \phi \end{bmatrix}}{\partial \begin{bmatrix} 0 & x & 0 & 0 \\ -x & 0 & 0 & 0 \\ 0 & 0 & 0 & -x \\ 0 & 0 & x & 0 \end{bmatrix}} \tag{12.10}$$

This is:

$$= \frac{\partial \begin{bmatrix} \phi & A_x & A_y & A_z \\ -A_x & \phi & -A_z & A_y \\ -A_y & A_z & \phi & -A_x \\ -A_z & -A_y & A_x & \phi \end{bmatrix}}{\begin{bmatrix} 0 & 1 & 0 & 0 \\ -1 & 0 & 0 & 0 \\ 0 & 0 & 0 & -1 \\ 0 & 0 & 1 & 0 \end{bmatrix} \frac{1}{\partial \begin{bmatrix} x & 0 & 0 & 0 \\ 0 & x & 0 & 0 \\ 0 & 0 & x & 0 \\ 0 & 0 & 0 & x \end{bmatrix}}} \tag{12.11}$$

$$= \begin{bmatrix} 0 & -1 & 0 & 0 \\ 1 & 0 & 0 & 0 \\ 0 & 0 & 0 & 1 \\ 0 & 0 & -1 & 0 \end{bmatrix} \begin{bmatrix} \dfrac{\partial \phi}{\partial x} & \dfrac{\partial A_x}{\partial x} & \dfrac{\partial A_y}{\partial x} & \dfrac{\partial A_z}{\partial x} \\[2mm] -\dfrac{\partial A_x}{\partial x} & \dfrac{\partial \phi}{\partial x} & -\dfrac{\partial A_z}{\partial x} & \dfrac{\partial A_y}{\partial x} \\[2mm] -\dfrac{\partial A_y}{\partial x} & \dfrac{\partial A_z}{\partial x} & \dfrac{\partial \phi}{\partial x} & -\dfrac{\partial A_x}{\partial x} \\[2mm] -\dfrac{\partial A_z}{\partial x} & -\dfrac{\partial A_y}{\partial x} & \dfrac{\partial A_x}{\partial x} & \dfrac{\partial \phi}{\partial x} \end{bmatrix} \tag{12.12}$$

$$= \begin{bmatrix} \dfrac{\partial A_x}{\partial x} & -\dfrac{\partial \phi}{\partial x} & \dfrac{\partial A_z}{\partial x} & -\dfrac{\partial A_y}{\partial x} \\[3mm] \dfrac{\partial \phi}{\partial x} & \dfrac{\partial A_x}{\partial x} & \dfrac{\partial A_y}{\partial x} & \dfrac{\partial A_z}{\partial x} \\[3mm] -\dfrac{\partial A_z}{\partial x} & -\dfrac{\partial A_y}{\partial x} & \dfrac{\partial A_x}{\partial x} & \dfrac{\partial \phi}{\partial x} \\[3mm] \dfrac{\partial A_y}{\partial x} & -\dfrac{\partial A_z}{\partial x} & -\dfrac{\partial \phi}{\partial x} & \dfrac{\partial A_x}{\partial x} \end{bmatrix} \tag{12.13}$$

The point is that we took the matrix from the denominator to the left; if we had taken the matrix from the denominator to the right, we would have a different result.

The reader might think that we ought to extract the imaginary matrices from the numerator as well as from the denominator so that we are differentiating only a real function with respect to a real variable. The reader would be correct, but, after we had taken the imaginary matrices out of the numerator, we would have to put them back into the numerator; the result would be the same as leaving them in, as we have done above.

We repeat the procedure above with the other two parts of the differential:

$$\dfrac{\partial \begin{bmatrix} \phi & A_x & A_y & A_z \\ -A_x & \phi & -A_z & A_y \\ -A_y & A_z & \phi & -A_x \\ -A_z & -A_y & A_x & \phi \end{bmatrix}}{\partial \begin{bmatrix} 0 & 0 & y & 0 \\ 0 & 0 & 0 & y \\ -y & 0 & 0 & 0 \\ 0 & -y & 0 & 0 \end{bmatrix}} \tag{12.14}$$

This is:

$$= \begin{bmatrix} 0 & 0 & -1 & 0 \\ 0 & 0 & 0 & -1 \\ 1 & 0 & 0 & 0 \\ 0 & 1 & 0 & 0 \end{bmatrix} \begin{bmatrix} \dfrac{\partial \phi}{\partial y} & \dfrac{\partial A_x}{\partial y} & \dfrac{\partial A_y}{\partial y} & \dfrac{\partial A_z}{\partial y} \\[2mm] -\dfrac{\partial A_x}{\partial y} & \dfrac{\partial \phi}{\partial y} & -\dfrac{\partial A_z}{\partial y} & \dfrac{\partial A_y}{\partial y} \\[2mm] -\dfrac{\partial A_y}{\partial y} & \dfrac{\partial A_z}{\partial y} & \dfrac{\partial \phi}{\partial y} & -\dfrac{\partial A_x}{\partial y} \\[2mm] -\dfrac{\partial A_z}{\partial y} & -\dfrac{\partial A_y}{\partial y} & \dfrac{\partial A_x}{\partial y} & \dfrac{\partial \phi}{\partial y} \end{bmatrix}$$

$$= \begin{bmatrix} \dfrac{\partial A_y}{\partial y} & -\dfrac{\partial A_z}{\partial y} & -\dfrac{\partial \phi}{\partial y} & \dfrac{\partial A_x}{\partial y} \\[2mm] \dfrac{\partial A_z}{\partial y} & \dfrac{\partial A_y}{\partial y} & -\dfrac{\partial A_x}{\partial y} & -\dfrac{\partial \phi}{\partial y} \\[2mm] \dfrac{\partial \phi}{\partial y} & \dfrac{\partial A_x}{\partial y} & \dfrac{\partial A_y}{\partial y} & \dfrac{\partial A_z}{\partial y} \\[2mm] -\dfrac{\partial A_x}{\partial y} & \dfrac{\partial \phi}{\partial y} & -\dfrac{\partial A_z}{\partial y} & \dfrac{\partial A_y}{\partial y} \end{bmatrix} \tag{12.15}$$

and finally:

$$\dfrac{\partial \begin{bmatrix} \phi & A_x & A_y & A_z \\ -A_x & \phi & -A_z & A_y \\ -A_y & A_z & \phi & -A_x \\ -A_z & -A_y & A_x & \phi \end{bmatrix}}{\partial \begin{bmatrix} 0 & 0 & 0 & z \\ 0 & 0 & -z & 0 \\ 0 & z & 0 & 0 \\ -z & 0 & 0 & 0 \end{bmatrix}} \tag{12.16}$$

This is:

$$= \begin{bmatrix} 0 & 0 & 0 & -1 \\ 0 & 0 & 1 & 0 \\ 0 & -1 & 0 & 0 \\ 1 & 0 & 0 & 0 \end{bmatrix} \begin{bmatrix} \dfrac{\partial \phi}{\partial z} & \dfrac{\partial A_x}{\partial z} & \dfrac{\partial A_y}{\partial z} & \dfrac{\partial A_z}{\partial z} \\ -\dfrac{\partial A_x}{\partial z} & \dfrac{\partial \phi}{\partial z} & -\dfrac{\partial A_z}{\partial z} & \dfrac{\partial A_y}{\partial z} \\ -\dfrac{\partial A_y}{\partial z} & \dfrac{\partial A_z}{\partial z} & \dfrac{\partial \phi}{\partial z} & -\dfrac{\partial A_x}{\partial z} \\ -\dfrac{\partial A_z}{\partial z} & -\dfrac{\partial A_y}{\partial z} & \dfrac{\partial A_x}{\partial z} & \dfrac{\partial \phi}{\partial z} \end{bmatrix}$$

$$= \begin{bmatrix} \dfrac{\partial A_z}{\partial z} & \dfrac{\partial A_y}{\partial z} & -\dfrac{\partial A_x}{\partial z} & -\dfrac{\partial \phi}{\partial z} \\ -\dfrac{\partial A_y}{\partial z} & \dfrac{\partial A_z}{\partial z} & \dfrac{\partial \phi}{\partial z} & -\dfrac{\partial A_x}{\partial z} \\ \dfrac{\partial A_x}{\partial z} & -\dfrac{\partial \phi}{\partial z} & \dfrac{\partial A_z}{\partial z} & -\dfrac{\partial A_y}{\partial z} \\ \dfrac{\partial \phi}{\partial z} & \dfrac{\partial A_x}{\partial z} & \dfrac{\partial A_y}{\partial z} & \dfrac{\partial A_z}{\partial z} \end{bmatrix} \quad (12.17)$$

We now have the left differential as the sum of these four parts. That left differential is:

$$d_L \left(\frac{\Phi_{Pot}}{Q_{(t,x,y,z)}} \right) \quad (12.18)$$

This is:

$$= \begin{bmatrix} \dfrac{\partial\phi}{\partial t} & \dfrac{\partial A_x}{\partial t} & \dfrac{\partial A_y}{\partial t} & \dfrac{\partial A_z}{\partial t} \\[2ex] -\dfrac{\partial A_x}{\partial t} & \dfrac{\partial\phi}{\partial t} & -\dfrac{\partial A_z}{\partial t} & \dfrac{\partial A_y}{\partial t} \\[2ex] -\dfrac{\partial A_y}{\partial t} & \dfrac{\partial A_z}{\partial t} & \dfrac{\partial\phi}{\partial t} & -\dfrac{\partial A_x}{\partial t} \\[2ex] -\dfrac{\partial A_z}{\partial t} & -\dfrac{\partial A_y}{\partial t} & \dfrac{\partial A_x}{\partial t} & \dfrac{\partial\phi}{\partial t} \end{bmatrix} + \begin{bmatrix} \dfrac{\partial A_x}{\partial x} & -\dfrac{\partial\phi}{\partial x} & \dfrac{\partial A_z}{\partial x} & -\dfrac{\partial A_y}{\partial x} \\[2ex] \dfrac{\partial\phi}{\partial x} & \dfrac{\partial A_x}{\partial x} & \dfrac{\partial A_y}{\partial x} & \dfrac{\partial A_z}{\partial x} \\[2ex] -\dfrac{\partial A_z}{\partial x} & -\dfrac{\partial A_y}{\partial x} & \dfrac{\partial A_x}{\partial x} & \dfrac{\partial\phi}{\partial x} \\[2ex] \dfrac{\partial A_y}{\partial x} & -\dfrac{\partial A_z}{\partial x} & -\dfrac{\partial\phi}{\partial x} & \dfrac{\partial A_x}{\partial x} \end{bmatrix}$$

$$+ \begin{bmatrix} \dfrac{\partial A_y}{\partial y} & -\dfrac{\partial A_z}{\partial y} & -\dfrac{\partial\phi}{\partial y} & \dfrac{\partial A_x}{\partial y} \\[2ex] \dfrac{\partial A_z}{\partial y} & \dfrac{\partial A_y}{\partial y} & -\dfrac{\partial A_x}{\partial y} & -\dfrac{\partial\phi}{\partial y} \\[2ex] \dfrac{\partial\phi}{\partial y} & \dfrac{\partial A_x}{\partial y} & \dfrac{\partial A_y}{\partial y} & \dfrac{\partial A_z}{\partial y} \\[2ex] -\dfrac{\partial A_x}{\partial y} & \dfrac{\partial\phi}{\partial y} & -\dfrac{\partial A_z}{\partial y} & \dfrac{\partial A_y}{\partial y} \end{bmatrix} + \begin{bmatrix} \dfrac{\partial A_z}{\partial z} & \dfrac{\partial A_y}{\partial z} & -\dfrac{\partial A_x}{\partial z} & -\dfrac{\partial\phi}{\partial z} \\[2ex] -\dfrac{\partial A_y}{\partial z} & \dfrac{\partial A_z}{\partial z} & \dfrac{\partial\phi}{\partial z} & -\dfrac{\partial A_x}{\partial z} \\[2ex] \dfrac{\partial A_x}{\partial z} & -\dfrac{\partial\phi}{\partial z} & \dfrac{\partial A_z}{\partial z} & -\dfrac{\partial A_y}{\partial z} \\[2ex] \dfrac{\partial\phi}{\partial z} & \dfrac{\partial A_x}{\partial z} & \dfrac{\partial A_y}{\partial z} & \dfrac{\partial A_z}{\partial z} \end{bmatrix} \quad (12.19)$$

For ease of presentation, we list only the top row of the resulting matrix. We have:

$$d_L\Phi_{Pot[1,1]} = \frac{\partial\phi}{\partial t} + \frac{\partial A_x}{\partial x} + \frac{\partial A_y}{\partial y} + \frac{\partial A_z}{\partial z}$$

$$d_L\Phi_{Pot[1,2]} = \left(\frac{\partial A_x}{\partial t} - \frac{\partial\phi}{\partial x}\right) + \left(\frac{\partial A_y}{\partial z} - \frac{\partial A_z}{\partial y}\right)$$

$$d_L\Phi_{Pot[1,3]} = \left(\frac{\partial A_y}{\partial t} - \frac{\partial\phi}{\partial y}\right) + \left(\frac{\partial A_z}{\partial x} - \frac{\partial A_x}{\partial z}\right) \quad (12.20)$$

$$d_L\Phi_{Pot[1,4]} = \left(\frac{\partial A_z}{\partial t} - \frac{\partial\phi}{\partial z}\right) + \left(\frac{\partial A_x}{\partial y} - \frac{\partial A_y}{\partial x}\right)$$

Wherein, we have rearranged the terms and included some brackets. The reader might see that we have what looks like a divergence and some curls in the above, (12.20).

Aside: Normally, the curl of a vector field is defined for only 3-dimensional vector fields. Although, technically, the divergence and the gradient are also defined for only 3-dimensional fields, it is obvious how to extend these definitions into higher dimensions; with the curl, this is not obvious. Indeed, the 3-dimensional curl is still used to conventionally define the electromagnetic field as a derivative of the Minkowski 4-potential. Basically, as with the vector cross-product, unless we have the matrix technique used here, we do not know how to form the curl derivatives of vector fields over 4-dimensional spaces.

Differentiation to the right:

We now differentiate to the right. This means that we extract the imaginary variable matrices from the denominator to the right rather than to the left. We have:

$$\cfrac{\partial \begin{bmatrix} \phi & A_x & A_y & A_z \\ -A_x & \phi & -A_z & A_y \\ -A_y & A_z & \phi & -A_x \\ -A_z & -A_y & A_x & \phi \end{bmatrix}}{\partial \begin{bmatrix} t & 0 & 0 & 0 \\ 0 & t & 0 & 0 \\ 0 & 0 & t & 0 \\ 0 & 0 & 0 & t \end{bmatrix}} = \begin{bmatrix} \dfrac{\partial \phi}{\partial t} & \dfrac{\partial A_x}{\partial t} & \dfrac{\partial A_y}{\partial t} & \dfrac{\partial A_z}{\partial t} \\ -\dfrac{\partial A_x}{\partial t} & \dfrac{\partial \phi}{\partial t} & -\dfrac{\partial A_z}{\partial t} & \dfrac{\partial A_y}{\partial t} \\ -\dfrac{\partial A_y}{\partial t} & \dfrac{\partial A_z}{\partial t} & \dfrac{\partial \phi}{\partial t} & -\dfrac{\partial A_x}{\partial t} \\ -\dfrac{\partial A_z}{\partial t} & -\dfrac{\partial A_y}{\partial t} & \dfrac{\partial A_x}{\partial t} & \dfrac{\partial \phi}{\partial t} \end{bmatrix} \quad (12.21)$$

This is the same as differentiation to the left; of course it is; the t matrix is just a real number, and real numbers commute with everything. The next step is different.

$$\frac{\partial \begin{bmatrix} \phi & A_x & A_y & A_z \\ -A_x & \phi & -A_z & A_y \\ -A_y & A_z & \phi & -A_x \\ -A_z & -A_y & A_x & \phi \end{bmatrix}}{\partial \begin{bmatrix} 0 & x & 0 & 0 \\ -x & 0 & 0 & 0 \\ 0 & 0 & 0 & -x \\ 0 & 0 & x & 0 \end{bmatrix}} \tag{12.22}$$

This is:

$$= \frac{\partial \begin{bmatrix} \phi & A_x & A_y & A_z \\ -A_x & \phi & -A_z & A_y \\ -A_y & A_z & \phi & -A_x \\ -A_z & -A_y & A_x & \phi \end{bmatrix}}{\partial \begin{bmatrix} x & 0 & 0 & 0 \\ 0 & x & 0 & 0 \\ 0 & 0 & x & 0 \\ 0 & 0 & 0 & x \end{bmatrix}} \cdot \frac{1}{\begin{bmatrix} 0 & 1 & 0 & 0 \\ -1 & 0 & 0 & 0 \\ 0 & 0 & 0 & -1 \\ 0 & 0 & 1 & 0 \end{bmatrix}} \tag{12.23}$$

$$= \begin{bmatrix} \dfrac{\partial \phi}{\partial x} & \dfrac{\partial A_x}{\partial x} & \dfrac{\partial A_y}{\partial x} & \dfrac{\partial A_z}{\partial x} \\ -\dfrac{\partial A_x}{\partial x} & \dfrac{\partial \phi}{\partial x} & -\dfrac{\partial A_z}{\partial x} & \dfrac{\partial A_y}{\partial x} \\ -\dfrac{\partial A_y}{\partial x} & \dfrac{\partial A_z}{\partial x} & \dfrac{\partial \phi}{\partial x} & -\dfrac{\partial A_x}{\partial x} \\ -\dfrac{\partial A_z}{\partial x} & -\dfrac{\partial A_y}{\partial x} & \dfrac{\partial A_x}{\partial x} & \dfrac{\partial \phi}{\partial x} \end{bmatrix} \begin{bmatrix} 0 & -1 & 0 & 0 \\ 1 & 0 & 0 & 0 \\ 0 & 0 & 0 & 1 \\ 0 & 0 & -1 & 0 \end{bmatrix} \tag{12.24}$$

$$= \begin{bmatrix} \dfrac{\partial A_x}{\partial x} & -\dfrac{\partial \phi}{\partial x} & -\dfrac{\partial A_z}{\partial x} & \dfrac{\partial A_y}{\partial x} \\[3mm] \dfrac{\partial \phi}{\partial x} & \dfrac{\partial A_x}{\partial x} & -\dfrac{\partial A_y}{\partial x} & -\dfrac{\partial A_z}{\partial x} \\[3mm] \dfrac{\partial A_z}{\partial x} & \dfrac{\partial A_y}{\partial x} & \dfrac{\partial A_x}{\partial x} & \dfrac{\partial \phi}{\partial x} \\[3mm] -\dfrac{\partial A_y}{\partial x} & \dfrac{\partial A_z}{\partial x} & -\dfrac{\partial \phi}{\partial x} & \dfrac{\partial A_x}{\partial x} \end{bmatrix} \tag{12.25}$$

We deal with the next two variables in the same way.

$$\partial \begin{bmatrix} \phi & A_x & A_y & A_z \\ -A_x & \phi & -A_z & A_y \\ -A_y & A_z & \phi & -A_x \\ -A_z & -A_y & A_x & \phi \end{bmatrix}$$
$$\overline{\partial \begin{bmatrix} 0 & 0 & y & 0 \\ 0 & 0 & 0 & y \\ -y & 0 & 0 & 0 \\ 0 & -y & 0 & 0 \end{bmatrix}} \tag{12.26}$$

This is:

$$= \begin{bmatrix} \dfrac{\partial \phi}{\partial y} & \dfrac{\partial A_x}{\partial y} & \dfrac{\partial A_y}{\partial y} & \dfrac{\partial A_z}{\partial y} \\[3mm] -\dfrac{\partial A_x}{\partial y} & \dfrac{\partial \phi}{\partial y} & -\dfrac{\partial A_z}{\partial y} & \dfrac{\partial A_y}{\partial y} \\[3mm] -\dfrac{\partial A_y}{\partial y} & \dfrac{\partial A_z}{\partial y} & \dfrac{\partial \phi}{\partial y} & -\dfrac{\partial A_x}{\partial y} \\[3mm] -\dfrac{\partial A_z}{\partial y} & -\dfrac{\partial A_y}{\partial y} & \dfrac{\partial A_x}{\partial y} & \dfrac{\partial \phi}{\partial y} \end{bmatrix} \begin{bmatrix} 0 & 0 & -1 & 0 \\ 0 & 0 & 0 & -1 \\ 1 & 0 & 0 & 0 \\ 0 & 1 & 0 & 0 \end{bmatrix} \tag{12.27}$$

$$= \begin{bmatrix} \dfrac{\partial A_y}{\partial y} & \dfrac{\partial A_z}{\partial y} & -\dfrac{\partial \phi}{\partial y} & -\dfrac{\partial A_x}{\partial y} \\[2ex] -\dfrac{\partial A_z}{\partial y} & \dfrac{\partial A_y}{\partial y} & \dfrac{\partial A_x}{\partial y} & -\dfrac{\partial \phi}{\partial y} \\[2ex] \dfrac{\partial \phi}{\partial y} & -\dfrac{\partial A_x}{\partial y} & \dfrac{\partial A_y}{\partial y} & -\dfrac{\partial A_z}{\partial y} \\[2ex] \dfrac{\partial A_x}{\partial y} & \dfrac{\partial \phi}{\partial y} & \dfrac{\partial A_z}{\partial y} & \dfrac{\partial A_y}{\partial y} \end{bmatrix} \qquad (12.28)$$

And:

$$\dfrac{\partial \begin{bmatrix} \phi & A_x & A_y & A_z \\ -A_x & \phi & -A_z & A_y \\ -A_y & A_z & \phi & -A_x \\ -A_z & -A_y & A_x & \phi \end{bmatrix}}{\partial \begin{bmatrix} 0 & 0 & 0 & z \\ 0 & 0 & -z & 0 \\ 0 & z & 0 & 0 \\ -z & 0 & 0 & 0 \end{bmatrix}} \qquad (12.29)$$

This is:

$$= \begin{bmatrix} \dfrac{\partial \phi}{\partial z} & \dfrac{\partial A_x}{\partial z} & \dfrac{\partial A_y}{\partial z} & \dfrac{\partial A_z}{\partial z} \\[2ex] -\dfrac{\partial A_x}{\partial z} & \dfrac{\partial \phi}{\partial z} & -\dfrac{\partial A_z}{\partial z} & \dfrac{\partial A_y}{\partial z} \\[2ex] \dfrac{\partial A_y}{\partial z} & \dfrac{\partial A_z}{\partial z} & \dfrac{\partial \phi}{\partial z} & -\dfrac{\partial A_x}{\partial z} \\[2ex] -\dfrac{\partial A_z}{\partial z} & -\dfrac{\partial A_y}{\partial z} & \dfrac{\partial A_x}{\partial z} & \dfrac{\partial \phi}{\partial z} \end{bmatrix} \begin{bmatrix} 0 & 0 & 0 & -1 \\ 0 & 0 & 1 & 0 \\ 0 & -1 & 0 & 0 \\ 1 & 0 & 0 & 0 \end{bmatrix} \qquad (12.30)$$

$$= \begin{bmatrix} \dfrac{\partial A_z}{\partial z} & -\dfrac{\partial A_y}{\partial z} & \dfrac{\partial A_x}{\partial z} & -\dfrac{\partial \phi}{\partial z} \\[2ex] \dfrac{\partial A_y}{\partial z} & \dfrac{\partial A_z}{\partial z} & \dfrac{\partial \phi}{\partial z} & \dfrac{\partial A_x}{\partial z} \\[2ex] -\dfrac{\partial A_x}{\partial z} & -\dfrac{\partial \phi}{\partial z} & \dfrac{\partial A_z}{\partial z} & \dfrac{\partial A_y}{\partial z} \\[2ex] \dfrac{\partial \phi}{\partial z} & -\dfrac{\partial A_x}{\partial z} & -\dfrac{\partial A_y}{\partial z} & \dfrac{\partial A_z}{\partial z} \end{bmatrix} \tag{12.31}$$

We now have the right differential as the sum of these four parts. That right differential is:

$$d_R \left(\dfrac{\Phi_{Pot}}{Q_{(t,x,y,z)}} \right) \tag{12.32}$$

$$= \begin{bmatrix} \dfrac{\partial \phi}{\partial t} & \dfrac{\partial A_x}{\partial t} & \dfrac{\partial A_y}{\partial t} & \dfrac{\partial A_z}{\partial t} \\[2ex] -\dfrac{\partial A_x}{\partial t} & \dfrac{\partial \phi}{\partial t} & -\dfrac{\partial A_z}{\partial t} & \dfrac{\partial A_y}{\partial t} \\[2ex] -\dfrac{\partial A_y}{\partial t} & \dfrac{\partial A_z}{\partial t} & \dfrac{\partial \phi}{\partial t} & -\dfrac{\partial A_x}{\partial t} \\[2ex] -\dfrac{\partial A_z}{\partial t} & -\dfrac{\partial A_y}{\partial t} & \dfrac{\partial A_x}{\partial t} & \dfrac{\partial \phi}{\partial t} \end{bmatrix} + \begin{bmatrix} \dfrac{\partial A_x}{\partial x} & -\dfrac{\partial \phi}{\partial x} & -\dfrac{\partial A_z}{\partial x} & \dfrac{\partial A_y}{\partial x} \\[2ex] \dfrac{\partial \phi}{\partial x} & \dfrac{\partial A_x}{\partial x} & \dfrac{\partial A_y}{\partial x} & -\dfrac{\partial A_z}{\partial x} \\[2ex] \dfrac{\partial A_z}{\partial x} & \dfrac{\partial A_y}{\partial x} & \dfrac{\partial A_x}{\partial x} & \dfrac{\partial \phi}{\partial x} \\[2ex] -\dfrac{\partial A_y}{\partial x} & \dfrac{\partial A_z}{\partial x} & -\dfrac{\partial \phi}{\partial x} & \dfrac{\partial A_x}{\partial x} \end{bmatrix}$$

$$+ \begin{bmatrix} \dfrac{\partial A_y}{\partial y} & \dfrac{\partial A_z}{\partial y} & -\dfrac{\partial \phi}{\partial y} & -\dfrac{\partial A_x}{\partial y} \\[2ex] -\dfrac{\partial A_z}{\partial y} & \dfrac{\partial A_y}{\partial y} & \dfrac{\partial A_x}{\partial y} & -\dfrac{\partial \phi}{\partial y} \\[2ex] \dfrac{\partial \phi}{\partial y} & -\dfrac{\partial A_x}{\partial y} & \dfrac{\partial A_y}{\partial y} & -\dfrac{\partial A_z}{\partial y} \\[2ex] \dfrac{\partial A_x}{\partial y} & \dfrac{\partial \phi}{\partial y} & \dfrac{\partial A_z}{\partial y} & \dfrac{\partial A_y}{\partial y} \end{bmatrix} + \begin{bmatrix} \dfrac{\partial A_z}{\partial z} & \dfrac{\partial A_y}{\partial z} & \dfrac{\partial A_x}{\partial z} & -\dfrac{\partial \phi}{\partial z} \\[2ex] \dfrac{\partial A_y}{\partial z} & \dfrac{\partial A_z}{\partial z} & \dfrac{\partial \phi}{\partial z} & \dfrac{\partial A_x}{\partial z} \\[2ex] -\dfrac{\partial A_x}{\partial z} & -\dfrac{\partial \phi}{\partial z} & \dfrac{\partial A_z}{\partial z} & \dfrac{\partial A_y}{\partial z} \\[2ex] \dfrac{\partial \phi}{\partial z} & -\dfrac{\partial A_x}{\partial z} & -\dfrac{\partial A_y}{\partial z} & \dfrac{\partial A_z}{\partial z} \end{bmatrix} \tag{12.33}$$

For ease of presentation, we list only the top row of the resulting matrix. We have:

$$d_R \Phi_{Pot[1,1]} = \frac{\partial \phi}{\partial t} + \frac{\partial A_x}{\partial x} + \frac{\partial A_y}{\partial y} + \frac{\partial A_z}{\partial z}$$

$$d_R \Phi_{Pot[1,2]} = \left(\frac{\partial A_x}{\partial t} - \frac{\partial \phi}{\partial x} \right) + \left(\frac{\partial A_z}{\partial y} - \frac{\partial A_y}{\partial z} \right)$$

$$d_R \Phi_{Pot[1,3]} = \left(\frac{\partial A_y}{\partial t} - \frac{\partial \phi}{\partial y} \right) + \left(\frac{\partial A_x}{\partial z} - \frac{\partial A_z}{\partial x} \right) \qquad (12.34)$$

$$d_R \Phi_{Pot[1,4]} = \left(\frac{\partial A_z}{\partial t} - \frac{\partial \phi}{\partial z} \right) + \left(\frac{\partial A_y}{\partial x} - \frac{\partial A_x}{\partial y} \right)$$

Wherein, we have rearranged the terms and included some brackets. The reader might see that we again have a divergence and some curls.

We now have both the left differential of the potential, d_L, and the right differential of the potential, d_R.

The E field and the B field:
We form the E field as:

$$E_{\mu\nu} = \frac{1}{2} \left(d_L + d_R \right) \qquad (12.35)$$

The E field is a quaternion matrix; of course it is; we are working within the quaternion algebra. For ease of presentation, we list only the top row of the E field matrix:

$$E_{[1,1]} = \frac{1}{2} \left(\frac{\partial \phi}{\partial t} + \frac{\partial A_x}{\partial x} + \frac{\partial A_y}{\partial y} + \frac{\partial A_z}{\partial z} + \frac{\partial \phi}{\partial t} + \frac{\partial A_x}{\partial x} + \frac{\partial A_y}{\partial y} + \frac{\partial A_z}{\partial z} \right)$$

$$(12.36)$$

$$E_{[1,2]} = \frac{1}{2} \left(\left(\frac{\partial A_x}{\partial t} - \frac{\partial \phi}{\partial x} \right) + \left(\frac{\partial A_y}{\partial z} - \frac{\partial A_z}{\partial y} \right) + \left(\frac{\partial A_x}{\partial t} - \frac{\partial \phi}{\partial x} \right) + \left(\frac{\partial A_z}{\partial y} - \frac{\partial A_y}{\partial z} \right) \right)$$

$$(12.37)$$

$$E_{[1,3]} = \frac{1}{2}\left(\left(\frac{\partial A_y}{\partial t} - \frac{\partial \phi}{\partial y}\right) + \left(\frac{\partial A_z}{\partial x} - \frac{\partial A_x}{\partial z}\right) + \left(\frac{\partial A_y}{\partial t} - \frac{\partial \phi}{\partial y}\right) + \left(\frac{\partial A_x}{\partial z} - \frac{\partial A_z}{\partial x}\right)\right)$$

(12.38)

$$E_{[1,4]} = \frac{1}{2}\left(\left(\frac{\partial A_z}{\partial t} - \frac{\partial \phi}{\partial z}\right) + \left(\frac{\partial A_x}{\partial y} - \frac{\partial A_y}{\partial x}\right) + \left(\frac{\partial A_z}{\partial t} - \frac{\partial \phi}{\partial z}\right) + \left(\frac{\partial A_y}{\partial x} - \frac{\partial A_x}{\partial y}\right)\right)$$

(12.39)

Leading to $E_{\mu\nu}$ as:

$$E_{[1,1]} = \frac{\partial \phi}{\partial t} + \frac{\partial A_x}{\partial x} + \frac{\partial A_y}{\partial y} + \frac{\partial A_z}{\partial z}$$

$$E_{[1,2]} = \left(\frac{\partial A_x}{\partial t} - \frac{\partial \phi}{\partial x}\right)$$

$$E_{[1,3]} = \left(\frac{\partial A_y}{\partial t} - \frac{\partial \phi}{\partial y}\right)$$

(12.40)

$$E_{[1,4]} = \left(\frac{\partial A_z}{\partial t} - \frac{\partial \phi}{\partial z}\right)$$

We form the B field as:

$$B_{\mu\nu} = \frac{1}{2}(d_L - d_R)$$

(12.41)

The B field is, of course, also a quaternion matrix. For ease of presentation, we list only the top row of the B field matrix.

$$B_{[1,1]} = \frac{1}{2}\left(\frac{\partial \phi}{\partial t} + \frac{\partial A_x}{\partial x} + \frac{\partial A_y}{\partial y} + \frac{\partial A_z}{\partial z} - \frac{\partial \phi}{\partial t} - \frac{\partial A_x}{\partial x} - \frac{\partial A_y}{\partial y} - \frac{\partial A_z}{\partial z}\right)$$

$$B_{[1,2]} = \frac{1}{2}\left(\left(\frac{\partial A_x}{\partial t} - \frac{\partial \phi}{\partial x}\right) + \left(\frac{\partial A_y}{\partial z} - \frac{\partial A_z}{\partial y}\right) - \left(\frac{\partial A_x}{\partial t} - \frac{\partial \phi}{\partial x}\right) - \left(\frac{\partial A_z}{\partial y} - \frac{\partial A_y}{\partial z}\right)\right)$$

$$B_{[1,3]} = \frac{1}{2}\left(\left(\frac{\partial A_y}{\partial t} - \frac{\partial \phi}{\partial y}\right) + \left(\frac{\partial A_z}{\partial x} - \frac{\partial A_x}{\partial z}\right) - \left(\frac{\partial A_y}{\partial t} - \frac{\partial \phi}{\partial y}\right) - \left(\frac{\partial A_x}{\partial z} - \frac{\partial A_z}{\partial x}\right)\right)$$

(12.42)

$$B_{[1,4]} = \frac{1}{2}\left(\left(\frac{\partial A_z}{\partial t} - \frac{\partial \phi}{\partial z}\right) + \left(\frac{\partial A_x}{\partial y} - \frac{\partial A_y}{\partial x}\right) - \left(\frac{\partial A_z}{\partial t} - \frac{\partial \phi}{\partial z}\right) - \left(\frac{\partial A_y}{\partial x} - \frac{\partial A_x}{\partial y}\right)\right)$$

Leading to $B_{\mu\nu}$ as:

$$B_{[1,1]} = 0$$

$$B_{[1,2]} = \left(\frac{\partial A_y}{\partial z} - \frac{\partial A_z}{\partial y} \right)$$

$$B_{[1,3]} = \left(\frac{\partial A_z}{\partial x} - \frac{\partial A_x}{\partial z} \right) \qquad (12.43)$$

$$B_{[1,4]} = \left(\frac{\partial A_x}{\partial y} - \frac{\partial A_y}{\partial x} \right)$$

Using the conventional definition of the curl of a 3-dimensional vector field, we could write this as:

$$\vec{B} = -Curl\left(\vec{A}\right) \qquad (12.44)$$

Quaternion curls, divergence, and gradient:
The reader will recognise the last three components of the B field of the quaternion potential as being the negative of the 3-dimensional curl. The direction of the 3-dimensional curl is arbitrary, and the existence of the $\overset{RL}{B}$ is consistent with this. We see that the non-commutative differentiation of a quaternion potential leads to the conventional curl up to sign. The last three components of the E field also has the conventional form of a curl:

$$\vec{E} = \frac{\partial \vec{A}}{\partial t} - grad\left(\phi\right) \qquad (12.45)$$

Aside: If we had differentiated the conjugate quaternion potential:

$$Q^{*}_{Pot} = \begin{bmatrix} \phi & -A_x & -A_y & -A_z \\ A_x & \phi & A_z & -A_y \\ A_y & -A_z & \phi & A_x \\ A_z & A_y & -A_x & \phi \end{bmatrix} \qquad (12.46)$$

With respect to a quaternion, we would have obtained:

$$\vec{E} = -\frac{\partial \vec{A}}{\partial t} - grad\left(\phi\right)$$

$$\vec{B} = Curl\left(\vec{A}\right)$$

(12.47)

This is the normal definition of the electric field and the magnetic field. Of course, the signs of these fields are arbitrary. We will see shortly that whether we choose to differentiate a quaternion or a conjugate quaternion leads to the same homogeneous Maxwell equations.

We have non-commutatively differentiated a quaternion potential with respect to a quaternion. The matrix forms of the eight $C_2 \times C_2$ algebras differ only in the distribution of minus signs, and so it is that the $E \& B$ fields of any one of the other seven algebras will differ from the above only by the distribution of minus signs.

Different curls for different algebras:
We assert that the B field is the 4-dimensional quaternion curl. We assert that $E_{[1,1]}$ is the 4-dimensional quaternion divergence. We assert that the other three components of the E field are the quaternion gradient. To avoid confusion, we perhaps ought to label the different curls etc. with the algebra to which they relate. Thus, we might occasionally use notation like:

$$Curl_Q = B_Q, \quad Curl_{Anti-Q} = B_{Anti-Q}, \quad Div_{SAS1} = E_{[1,1]SAS1} \quad (12.48)$$

In general, when the context is clear, and we are working within a single algebra, we will avoid such cumbersome notation. The point is that each algebra has its own type of curl although they often differ by no more than a minus sign.

Summary:
Well! we have produced curls, divergences and gradients. We have, if we choose the conjugate quaternion potential, arrived at the

conventional definitions of the electric field and the magnetic field except that our definitions include within them the $SU(2)$ commutation relations that are intrinsic to the quaternions. We have not started from 'obvious' assumptions and logically deduced that non-commutative differentiation is obviously the correct form of differentiation within a non-commutative division algebra. We have pulled non-commutative differentiation out of thin air, and we now assert that it is a correct mathematical procedure because it gives the 'correct' results. These results are not contrived, and they do seem to be 'correct', but we still have no deep understanding of why non-commutative differentiation is the 'correct' differentiation[52].

It might be that differentiation within a division algebra is associated with rotation. Division algebras are associated with rotation as is seen in their polar forms. The trigonometric functions of a space (think complex plane $\mathbb{C}$ and the $\{\cos(\ ), \sin(\ )\}$ functions) are projections on to particular axes. The trigonometric functions differentiate into each other, and so we can view differentiation within a division algebra as rotation from one axis to another axis (through 90° in $\mathbb{C}$). We will see later, that rotation in non-commutative spaces necessarily involves the commutator and anti-commutators; this is because rotation in non-commutative spaces through an angle, say, of (b,c,d) followed by rotation through an angle, say, of (f,g,h) is not a rotation through $(b+f,\ c+g,\ d+h)$ as it would be if the space was commutative.

Is non-commutative differentiation a sensible differentiation operation? That is for the reader to decide.

This has been a cumbersome chapter, but we have demonstrated, clearly, we hope, how to form a non-commutative differential. Such differentiation does not work with matrices in general. We have shown that it does work within a division algebra. It also works between the different algebras of the $C_2 \times C_2$ group; this is because

---

[52] It matches the accepted form of differentiation within Clifford algebras.

they share the same basic form of variables – symmetric and anti-symmetric. Whether or not such inter-algebra differentiation is meaningful is, as yet, an unresolved question, but we point out the physicists routinely differentiate electromagnetic fields with respect to space-time variables; is this not inter-algebra differentiation?

The reader doubtless feels that she would not wish to do the cumbersome calculations above every second of every hour of every day of the remainder of her life. Such calculations can be done by computer. Otherwise, we will later present an easier, but obscure, way of doing the calculations by thinking of non-commutative differentiation as a $SU(2)$ differential operator.

Chapter 13

Non-commutative Double Differentiation

In the previous chapter, we defined the left differential, d_L, and the right differential, d_R of a quaternion potential, Φ_{Pot}, with respect to a quaternion variable, $Q_{(t,x,y,z)}$. We defined the E field and the B field as:

$$E_{\mu\nu} = \frac{1}{2}\left(d_L + d_R\right)$$

$$B_{\mu\nu} = \frac{1}{2}\left(d_L - d_R\right)$$

(13.1)

Of course, both $\{d_R, d_L\}$ are quaternion matrices. (Everything is a quaternion matrix when we do algebraic operations within the quaternion algebra.) Since $\{d_R, d_L\}$ are quaternion matrices, we can differentiate each of them to the left or to the right. This will lead to four double differentials. To keep the notation unambiguous, we will look upon $\{d_R, d_L\}$ as being differential operators that act upon the object within the curved brackets that follows them. We will continue to omit the $\partial Q_{(t,x,y,z)}$ part. We thus have the right differential of the quaternion potential written as $d_R(\Phi_{Pot})$ and the left differential of the quaternion potential written as $d_L(\Phi_{Pot})$. The right and left differentials of these first differentials are then written as:

$$d_R\left(d_R\left(\Phi_{Pot}\right)\right) \equiv d_R d_R \Phi_{Pot}$$

$$d_R\left(d_L\left(\Phi_{Pot}\right)\right) \equiv d_R d_L \Phi_{Pot}$$

(13.2)

$$d_L\big(d_R(\Phi_{Pot})\big) \equiv d_L d_R \Phi_{Pot}$$
$$d_L\big(d_L(\Phi_{Pot})\big) \equiv d_L d_L \Phi_{Pot}$$

$$(13.3)$$

Wherein, having made clear what we mean, we have abbreviated the notation again by removing the brackets.

First differential identity:

Because matrices are associative, $(AB)C = A(BC)$, or by cumbersome calculation, it follows that:

$$d_L d_R (\Phi_{Pot}) = d_R d_L (\Phi_{Pot})$$

$$(13.4)$$

This is true in general of any non-commutative $C_2 \times C_2$ algebraic matrix not only Φ_{Pot}.

Other differential identities:

Henceforward, to avoid cluttering the page, we will omit the $\mu\nu$ subscript from the $E \& B$ fields. We have:

$$d_R E = d_R \frac{1}{2}\big(d_L(\Phi_{Pot}) + d_R(\Phi_{Pot})\big)$$
$$= \frac{1}{2}\big(d_R\big(d_L(\Phi_{Pot})\big) + d_R\big(d_R(\Phi_{Pot})\big)\big)$$
$$= \frac{1}{2}\big(d_R d_L + d_R d_R\big)$$

$$(13.5)$$

$$d_L E = d_L \frac{1}{2}\big(d_L(\Phi_{Pot}) + d_R(\Phi_{Pot})\big)$$
$$= \frac{1}{2}\big(d_L\big(d_L(\Phi_{Pot})\big) + d_L\big(d_R(\Phi_{Pot})\big)\big)$$
$$= \frac{1}{2}\big(d_L d_L + d_L d_R\big)$$

$$(13.6)$$

We also have:

139

$$d_R B = d_R \frac{1}{2}\left(d_L\left(\Phi_{Pot}\right) - d_R\left(\Phi_{Pot}\right)\right)$$

$$= \frac{1}{2}\left(d_R\left(d_L\left(\Phi_{Pot}\right)\right) - d_R\left(d_R\left(\Phi_{Pot}\right)\right)\right) \qquad (13.7)$$

$$= \frac{1}{2}\left(d_R d_L - d_R d_R\right)$$

$$d_L B = d_L \frac{1}{2}\left(d_L\left(\Phi_{Pot}\right) - d_R\left(\Phi_{Pot}\right)\right)$$

$$= \frac{1}{2}\left(d_L\left(d_L\left(\Phi_{Pot}\right)\right) - d_L\left(d_R\left(\Phi_{Pot}\right)\right)\right) \qquad (13.8)$$

$$= \frac{1}{2}\left(d_L d_L - d_L d_R\right)$$

We deal with these second differentials as we dealt with the first differentials by forming half the sum and half the difference of them. In doing this, we follow the convention that we subtract the right differential from the left differential. Not following this convention can lead to confusion[53]. We signify these sums and differences with { } and [ ] brackets respectively. We have:

$$\{E,d\} = \frac{1}{2}\left(d_L E + d_R E\right)$$

$$= \frac{1}{4}\left(\left(d_R d_L + d_R d_R\right) + \left(d_L d_L + d_L d_R\right)\right)\Phi_{Pot} \qquad (13.9)$$

$$= \frac{1}{4}\left(d_L d_L + 2d_L d_R + d_R d_R\right)\Phi_{Pot}$$

We call this curly-E.

We have:

---

[53] Hm! I wonder how your author discovered this surprising fact.

$$[E,d] = \frac{1}{2}(d_L E - d_R E)$$

$$= \frac{1}{4}\left((d_L d_L + d_L d_R) - (d_R d_L + d_R d_R)\right)\Phi_{Pot} \qquad (13.10)$$

$$= \frac{1}{4}(d_L d_L - d_R d_R)\Phi_{Pot}$$

We call this straight-E.

We have:

$$\{B,d\} = \frac{1}{2}(d_L B + d_R B)$$

$$= \frac{1}{4}\left((d_L d_L - d_L d_R) + (d_R d_L - d_R d_R)\right)\Phi_{Pot} \qquad (13.11)$$

$$= \frac{1}{4}(d_L d_L - d_R d_R)\Phi_{Pot}$$

We call this curly-B.

We have:

$$[B,d] = \frac{1}{2}(d_L B - d_R B)$$

$$= \frac{1}{4}\left((d_L d_L - d_L d_R) - (d_R d_L - d_R d_R)\right)\Phi_{Pot} \qquad (13.12)$$

$$= \frac{1}{4}(d_L d_L - 2d_L d_R + d_R d_R)\Phi_{Pot}$$

We call this straight-B.

We have a quite important differential identity:

$$\{B,d\} = [E,d] \qquad (13.13)$$

The components of this identity are the homogeneous Maxwell equations of the algebra[54]. Within quaternions, these are the no-magnetic monopoles law, $Div(\vec{B}) = 0$, and the Faraday equations of electromagnetism.

We have not, at this stage, derived the inhomogeneous Maxwell equations. The inhomogeneous Maxwell equations 'fall out' when the quaternion algebras are aggregated together.

Differential identities:

$$d_L d_R = d_R d_L$$
$$\{B,d\} = [E,d]$$

(13.14)

These differential identities are based upon differentiating with respect to the same matrix at every stage. If we were to differentiate with respect to a quaternion at the first stage and then with respect to a conjugate quaternion at the second stage, these identities would not apply.

Although we used the quaternions to demonstrate the above differential identities, the above differential identities are true of any of the non-commutative $C_2 \times C_2$ algebras. So, not only does each algebra have its own curl, divergence and gradient, but also its own set of homogeneous Maxwell equations, which are quantitised with the appropriate commutation relations of the algebra.

---

[54] Since this differential identity is the electromagnetic Maxwell equations within the quaternion algebra, we refer to such sets of similarly derived equations in other algebras as the Maxwell equations of that algebra. Clearly, James Clerk Maxwell is not involved, but we hope he would not mind our usage of his name.

Chapter 14

A Rewrite of Electromagnetism

In this chapter, electromagnetism is rewritten using quaternions. This formulation of electromagnetism has intrinsic $SU(2)$ commutation relations, and so we do not need to arbitrarily 'quantitise' the equations of motion by imposing commutation relations upon them in the way we normally do when we form QED. We use the 4-dimensional version of the curl operation which we introduced above as non-commutative differentiation. This 4-dimensional curl produces eight components of the electric and magnetic fields rather than the six components produced by the conventional curl of the 4-potential that are the conventional electromagnetic 4-tensor. The extra two components are the electric and magnetic fields in the time direction. The other results of the rewrite concur with standard theory. We introduce the $SU(2)$ differential operator (the quaternion differential operator). In the next chapter, we will rewrite electromagnetism again when we derive electromagnetism as the conventional 'unquantitised' electromagnetic tensor.

The potential:
Conventionally, in electromagnetism, we postulate a 4-potential of the form:

$$\Phi = \left[\phi, -A_x, -A_y, -A_z\right] \tag{14.1}$$

It is often not realised that, without being silly, there are three other possible potentials that may be assumed and that lead to a consistent presentation of electromagnetism. These potentials are:

$$\Phi = \left[\phi, A_x, A_y, A_z\right] \tag{14.2}$$

$$\Phi_{Anti} = \left[-\phi, -A_x, -A_y, -A_z\right]$$
$$\Phi_{Anti} = \left[-\phi, A_x, A_y, A_z\right]$$

(14.3)

The last two of these give rise to an anti-matter version of electromagnetism.

The four potentials given above are matched within the quaternions. Using quaternions, we may postulate a potential that is either a quaternion or a conjugate quaternion, and we may choose the real part of the quaternion or conjugate quaternion to be positive or negative. For presentational ease, we choose the conjugate quaternion potential because this gives definitions of the electric and magnetic fields which concur with the conventional definitions. The quaternion potential works perfectly well, as would the 4-potential without the minus signs, but this gives different definitions of the electric and magnetic fields and the results although correct are unfamiliar. Our conjugate quaternion potential is:

$$\Phi_{Q^*} = \begin{bmatrix} \phi & -A_x & -A_y & -A_z \\ A_x & \phi & A_z & -A_y \\ A_y & -A_z & \phi & A_x \\ A_z & A_y & -A_x & \phi \end{bmatrix}$$

(14.4)

Within this work, your author would have liked to have been able to claim that he assumes no more than the existence of the real numbers and the finite groups and that from only these assumptions we derive electromagnetism. However, we see here that we have assumed the existence of a potential. What is it to assume a potential? This potential is no more than a quaternion, but we assume that the variables within this quaternion vary from point to point over space-time. We then differentiate this quaternion with respect to those space-time variables and, as we will show shortly, we call the differentials, the anti-commutator and commutator of the rates at which this quaternion varies over space time, the electric and magnetic fields. We associate an acceleration through space-time with these electromagnetic fields in the form of a force:

$$\vec{E} = \frac{\vec{F}}{q} \tag{14.5}$$

We see, ignoring the charge, q, for now, that the electromagnetic forces are the anti-commutator and the commutator of a rate of change of a quaternion with respect to space-time, and so, when we assume a potential, we are assuming that a quaternion can change from point to point over space-time and that such a change is associated with a force.

Aside: In QFT, we allow the phase (angle) of the Lie algebras, $U(1)$ or $SU(2)$ to vary locally over space-time. This gives rise to potential terms. If we use the quaternion algebra instead of $SU(2)$ - they have the same commutation relations – and we identify local angle variation with variation of the imaginary variables and we also allow variation of the real variable, then we have a quaternion potential. So, when we assume a quaternion potential that varies over space-time, we are doing only what is done in QFT. Hm! perhaps we need to hold the real variable fixed to provide a basic scale to everything.

We have no idea why the anti-commutator and the commutator of such a change in a quaternion over space-time should manifest themselves as forces, but it seems that we do now understand the nature of Faraday's force fields a little more profoundly. A force is associated with the anti-commutator and commutator of the rate of change of one algebra with respect to the variables of another space (not necessarily a division algebra).

I what follows, we will differentiate Φ_{Q^*} with respect to the space-time variables $\{t, x, y, z\}$ presented as the quaternion matrix variables rather than as, for example, the SAS matrix variables or the matrix variables of the $SUM(A_3)$ matrix. If we are going to act within a particular algebra, we must do the differentiation this way. It works, and so we are left to ponder upon what we are really doing.

The conventional curl and the e/magnetic 4-tensor:

Conventionally, having defined the 4-potential as (14.1), the electromagnetic tensor is formed as the set of curls of this potential:

$$F_{\mu\nu} = \frac{\partial \Phi_\nu}{\partial x_\mu} - \frac{\partial \Phi_\mu}{\partial x_\nu} \tag{14.6}$$

Taking account of the minus signs in the potential, this gives

$$F_{12} = -\frac{\partial A_x}{\partial t} - \frac{\partial \phi}{\partial x} = E_x \qquad F_{43} = \left(\frac{\partial A_z}{\partial y} - \frac{\partial A_y}{\partial z} \right) = B_x$$

$$F_{13} = -\frac{\partial A_y}{\partial t} - \frac{\partial \phi}{\partial y} = E_y \qquad : \qquad F_{24} = \left(\frac{\partial A_x}{\partial z} - \frac{\partial A_z}{\partial x} \right) = B_y \tag{14.7}$$

$$F_{14} = -\frac{\partial A_z}{\partial t} - \frac{\partial \phi}{\partial z} = E_z \qquad F_{32} = \left(\frac{\partial A_y}{\partial x} - \frac{\partial A_x}{\partial y} \right) = B_z$$

Two fields, the electric field and the magnetic field, are assigned to these components. The directions of the fields (plus or minus) is arbitrary, but we have given the standard choice. With the above assignments, we see that the electric and magnetic fields are effectively defined as:

$$\vec{B} = Curl\left(\vec{A}\right)$$

$$\vec{E} = -Grad\left(\phi\right) - \frac{\partial \vec{A}}{\partial t} \tag{14.8}$$

There is a point here that is seldom mentioned. We accept that we live in a 4-dimensional space-time, but the traditional curl that we use in 3-dimensional space is used to act upon a 4-potential. Because the curl, $F_{\mu\nu}$, is zero when $\mu = \nu$, we derive only six field components when we ought to have eight field components; the $E_t \ \& \ B_t$ components are missing. We surely ought to act upon the potential with a 4-dimensional version of the curl.

The $SU(2)$ differential operator:

We are going to act upon a conjugate quaternion potential written as a 4×4 matrix. We will act upon this matrix by matrix multiplication with the quaternion differential operator. (We might also refer to this as the $SU(2)$ differential operator):

$$\partial Q = \begin{bmatrix} \partial t & -\partial x & -\partial y & -\partial z \\ \partial x & \partial t & \partial z & -\partial y \\ \partial y & -\partial z & \partial t & \partial x \\ \partial z & \partial y & -\partial x & \partial t \end{bmatrix} \qquad (14.9)$$

Wherein we have used $\partial x_i \equiv \dfrac{\partial}{\partial x_i}$. Note that the $SU(2)$ differential operator appears as the conjugate of the normal quaternion matrix[55], but this is deceptive; the differential operator of a division algebra has the format of the inverses of the individual variables; the inverse of an anti-symmetric imaginary variable is its negative; the inverses of symmetric imaginary variables (square roots of plus unity) are themselves.

The $SU(2)$ differential operator acts as if by matrix multiplication.

Applying the $SU(2)$ differential operator to a quaternion gives the same results as quaternion non-commutative differentiation. The quaternions are non-commutative, and so we get different results when we differentiate the potential from the left or differentiate the potential from the right. We refer to these results as the left differential and the right differential and denote them by $\{d_L \ \& \ d_R\}$:

---

[55] The $U(1)$ differential operator (the quantum mechanical momentum operator) is also a conjugate.

$$d_L\left(\Phi_{Q^*}\right) = \partial Q \Phi_{Q^*} =$$

$$\begin{bmatrix} \partial t & -\partial x & -\partial y & -\partial z \\ \partial x & \partial t & \partial z & -\partial y \\ \partial y & -\partial z & \partial t & \partial x \\ \partial z & \partial y & -\partial x & \partial t \end{bmatrix} \begin{bmatrix} \phi & -A_x & -A_y & -A_z \\ A_x & \phi & A_z & -A_y \\ A_y & -A_z & \phi & A_x \\ A_z & A_y & -A_x & \phi \end{bmatrix}$$

(14.10)

And:

$$d_R\left(\Phi_{Q^*}\right) = \Phi_{Q^*} \partial Q =$$

$$\begin{bmatrix} \phi & -A_x & -A_y & -A_z \\ A_x & \phi & A_z & -A_y \\ A_y & -A_z & \phi & A_x \\ A_z & A_y & -A_x & \phi \end{bmatrix} \begin{bmatrix} \partial t & -\partial x & -\partial y & -\partial z \\ \partial x & \partial t & \partial z & -\partial y \\ \partial y & -\partial z & \partial t & \partial x \\ \partial z & \partial y & -\partial x & \partial t \end{bmatrix}$$

(14.11)

We see that we are effectively doing non-commutative differentiation using a 'shortcut' which we call the $SU(2)$ differential operator. The use of the operator obfuscates what is really happening, but it uses less paper – I hope it does not catch on.

Having calculated the left and right differentials, as above, we form the $B \& E$ fields as:

$$E = \frac{1}{2}\left(d_L + d_R\right)$$

$$B = \frac{1}{2}\left(d_L - d_R\right)$$

(14.12)

Wherein we have omitted the Φ_{Q^*} for brevity.

Because we have only limited paper, we will present only the top row of the differentiated matrices. (The other components of the matrix are the same, other than sign, because we are working within a division algebra.) From (14.10) we get:

$$d_{L[1,1]} = \frac{\partial \phi}{\partial t} - \frac{\partial A_x}{\partial x} - \frac{\partial A_y}{\partial y} - \frac{\partial A_z}{\partial z}$$

$$d_{L[1,2]} = \left(-\frac{\partial \phi}{\partial x} - \frac{\partial A_x}{\partial t} \right) + \left(-\frac{\partial A_y}{\partial z} + \frac{\partial A_z}{\partial y} \right)$$

$$d_{L[1,3]} = \left(-\frac{\partial \phi}{\partial y} - \frac{\partial A_y}{\partial t} \right) + \left(\frac{\partial A_x}{\partial z} - \frac{\partial A_z}{\partial x} \right) \qquad (14.13)$$

$$d_{L[1,4]} = \left(-\frac{\partial \phi}{\partial z} - \frac{\partial A_z}{\partial t} \right) + \left(-\frac{\partial A_x}{\partial y} + \frac{\partial A_y}{\partial x} \right)$$

and from (14.11):

$$d_{R[1,1]} = \frac{\partial \phi}{\partial t} - \frac{\partial A_x}{\partial x} - \frac{\partial A_y}{\partial y} - \frac{\partial A_z}{\partial z}$$

$$d_{R[1,2]} = \left(-\frac{\partial \phi}{\partial x} - \frac{\partial A_x}{\partial t} \right) + \left(\frac{\partial A_y}{\partial z} - \frac{\partial A_z}{\partial y} \right)$$

$$d_{R[1,3]} = \left(-\frac{\partial \phi}{\partial y} - \frac{\partial A_y}{\partial t} \right) + \left(-\frac{\partial A_x}{\partial z} + \frac{\partial A_z}{\partial x} \right) \qquad (14.14)$$

$$d_{R[1,4]} = \left(-\frac{\partial \phi}{\partial z} - \frac{\partial A_z}{\partial t} \right) + \left(\frac{\partial A_x}{\partial y} - \frac{\partial A_y}{\partial x} \right)$$

This leads, (14.12), to the 4×4 E field quaternion matrix:

$$E_t = \frac{\partial \phi}{\partial t} - \frac{\partial A_x}{\partial x} - \frac{\partial A_y}{\partial y} - \frac{\partial A_z}{\partial z}$$

$$E_x = -\frac{\partial \phi}{\partial x} - \frac{\partial A_x}{\partial t}$$

$$E_y = -\frac{\partial \phi}{\partial y} - \frac{\partial A_y}{\partial t} \qquad (14.15)$$

$$E_z = -\frac{\partial \phi}{\partial z} - \frac{\partial A_z}{\partial t}$$

Wherein we have taken the real part of the quaternion, $E_{[1,1]}$ to be the time component and the three imaginary parts $\{E_{[1,2]}, E_{[1,3]}, E_{[1,4]}\}$ to be the spatial components.

The 4×4 B field quaternion matrix is, (14.12):

$$B_t = 0$$

$$B_x = \frac{\partial A_z}{\partial y} - \frac{\partial A_y}{\partial z}$$

$$B_y = \frac{\partial A_x}{\partial z} - \frac{\partial A_z}{\partial x} \qquad (14.16)$$

$$B_z = \frac{\partial A_y}{\partial x} - \frac{\partial A_x}{\partial y}$$

The reader will see that the six spatial parts of the $B \& E$ fields correspond to the six spatial parts of the magnetic and electric fields as were defined using the 4-potential and the conventional curl in the form of the electromagnetic tensor, (14.7). We also have the $B_t \& E_t$ components.

We think of the B field and the E field together as the 4-dimensional version of the 3-dimensional vector differentials curl/divergence/gradient[56].

Aside: There is something deceptive here that is obscured by the mathematics. The electric field is a 4-dimensional field; it simultaneously has components in all four directions within space-time. The magnetic field is a 2-dimensional field. There are no magnetic monopoles and the magnetic field arises from the motion of a charged particle. Such a field has components in only the two

---

[56] The same operation applies to the other seven non-commutative $C_2 \times C_2$ algebras, and so each of these algebras has its own version of the $B \& E$ fields. Similar operations apply within all non-commutative division algebras; of particular interest to physicists might be the 8-dimensional algebras of the $C_2 \times C_2 \times C_2$ group and the 16-dimensional algebras of the $C_2 \times C_2 \times C_2 \times C_2$ group.

directions that are spatially perpendicular to the direction of the velocity of the charged particle. If we consider the superposition of the magnetic fields of several charged particles moving in different directions, we have what appears to be a 3-dimensional magnetic field – such is a bar magnet. None-the-less, our '3-dimensional magnetic field' is no more than several 2-dimensional magnetic fields taken together. Why should we have a 2-dimensional magnetic field rather than a 4-dimensional one? No-one knows, but, if we calculate the $B \& E$ fields of an 8-dimensional $C_2 \times C_2 \times C_2$ algebra, we get an E field with eight components of which four are the 4-dimensional electric field and a B field with six non-zero components of which only two are the '4-dimensional' magnetic field. Note that two of the eight E field components are 'divergences' leaving six 'spatial' E field components to match the six B field components – perhaps this is something to do with the strong force.

Differential identities:
We remind the reader of the differential identity:

$$\{B,d\} = [E,d] \tag{14.17}$$

This unconditional differential identity is the four homogeneous Maxwell equations.

The homogeneous Maxwell equations:
Conventionally, having defined the electric and magnetic field as (14.8), the three spatial homogeneous Maxwell equations follow from these definitions as a differential identity (write them as double differentials of the potential). Within the quaternion algebra, we have the differential identity (14.17). For a conjugate quaternion potential differentiated with respect to a quaternion, this differential identity is the standard Faraday law and the no magnetic monopoles law (we give only the top row of the quaternion matrix equation):

$$\{B,d\} = [E,d]$$

$$\frac{\partial B_x}{\partial x} + \frac{\partial B_y}{\partial y} + \frac{\partial B_z}{\partial z} = 0$$

$$\frac{\partial B_x}{\partial t} = \frac{\partial E_y}{\partial z} - \frac{\partial E_z}{\partial y}$$

$$\frac{\partial B_y}{\partial t} = \frac{\partial E_z}{\partial x} - \frac{\partial E_x}{\partial z}$$

(14.18)

$$\frac{\partial B_z}{\partial t} = \frac{\partial E_x}{\partial y} - \frac{\partial E_y}{\partial x}$$

Either conventionally or quaternion wise, the homogeneous Maxwell equations are no more than differential identities. Conventionally the homogeneous Maxwell equations come as a set of three equations and another unconnected equation; quaternion wise, they come as one set of four equations - neater[57].

Lorentz transformations of the fields:
A Lorentz transformation, also referred to as a boost, is a 2-dimensional rotation in a space-time plane. It is accomplished in the (t,x) plane by the 2×2 space-time rotation matrix:

$$\begin{bmatrix} \cosh\chi & \sinh\chi \\ \sinh\chi & \cosh\chi \end{bmatrix}_{tx} \equiv \begin{bmatrix} \gamma_x & v_x\gamma_x \\ v_x\gamma_x & \gamma_x \end{bmatrix}$$

(14.19)

We also have 'anti-clockwise' rotation in the opposite direction:

$$\begin{bmatrix} \cosh\chi & -\sinh\chi \\ -\sinh\chi & \cosh\chi \end{bmatrix}_{tx} \equiv \begin{bmatrix} \gamma_x & -v_x\gamma_x \\ -v_x\gamma_x & \gamma_x \end{bmatrix}$$

(14.20)

Conventionally, we act separately with these 2-dimensional rotation matrices upon each component of the assumed 4-potential and thereby calculate how these components are changed by a velocity

---

[57] Anecdotally, Einstein was once told that an untidy desk was indicative of an untidy mind. He responded by asking what an empty desk was indicative of.

boost. From these boosted components, we form the boosted electromagnetic fields as:

$$E_x' = E_x \quad : \quad B_x' = B_x$$

$$E_y' = \gamma\left(E_y - v_x B_z\right) \quad : \quad B_y' = \gamma\left(B_y + v_x E_z\right) \qquad (14.21)$$

$$E_z' = \gamma\left(E_z + v_x B_y\right) \quad : \quad B_z' = \gamma\left(B_z - v_x E_y\right)$$

We will do such a calculation shortly. These transformations can also be written as:

$$\begin{bmatrix} \cosh \chi & -\sinh \chi \\ -\sinh \chi & \cosh \chi \end{bmatrix}_{tx} \begin{bmatrix} E_y & B_z \\ B_z & E_y \end{bmatrix}$$

$$= \cosh \chi_{tx} \begin{bmatrix} E_y - v_x B_z & B_z - v_x E_y \\ B_z - v_x E_y & E_y - v_x B_z \end{bmatrix} \qquad (14.22)$$

And:

$$\begin{bmatrix} \cosh \chi & \sinh \chi \\ \sinh \chi & \cosh \chi \end{bmatrix}_{tx} \begin{bmatrix} E_z & B_y \\ B_y & E_z \end{bmatrix}$$

$$= \cosh \chi_{tx} \begin{bmatrix} E_z + v_x B_y & B_y + v_x E_z \\ B_y + v_x E_z & E_z + v_x B_y \end{bmatrix} \qquad (14.23)$$

We know that, at least the $\left\{E_y, E_z, B_y, B_z\right\}$ transformations are correct because they lead directly to the observationally verified Lorentz force laws[58]. Similarly, for boosts in the $\left\{(t,y),(t,z)\right\}$ planes, we have transformations that lead to the Lorentz force laws. However, there is much glossed over in the above conventional presentation.

---

[58] Your author knows of no observational evidence to verify the transformations in the directions of the boost, but this might be no more than your author's ignorance.

Conventional Lorentz transform of e/mag fields:

Although it is not often stated in conventional presentations, we begin with the assumption that we live in six separate 2-dimensional spaces. Without this assumption, the potential in the other two directions would not be unaffected by the transformation. One of these 2-dimensional spaces is the (t, x) plane, and it is with rotation in that plane that we are concerned. What we calculate in this case would result equally well from choosing a different space-time plane.

The Lorentz transformation in the (t, x) plane of the 4-potential is:

$$
\begin{bmatrix} \phi & -A_x \\ -A_x & \phi \end{bmatrix}' = \begin{bmatrix} \gamma_x & v_x\gamma_x \\ v_x\gamma_x & \gamma_x \end{bmatrix} \begin{bmatrix} \phi & -A_x \\ -A_x & \phi \end{bmatrix}
$$
$$
= \begin{bmatrix} \gamma_x\phi - v_x\gamma_x A_x & -\gamma_x A_x + v_x\gamma_x\phi \\ -\gamma_x A_x + v_x\gamma_x\phi & \gamma_x\phi - v_x\gamma_x A_x \end{bmatrix}
$$

(14.24)

We note that these two matrices are commutative and it matters not whether the rotation is to the left or to the right. Using the 'six separate 2-dimensional spaces' assumption, we get:

$$
\begin{bmatrix} 0 & -A_y \\ -A_y & 0 \end{bmatrix}' = \begin{bmatrix} 0 & -A_y \\ -A_y & 0 \end{bmatrix}
$$
$$
\begin{bmatrix} 0 & -A_z \\ -A_z & 0 \end{bmatrix}' = \begin{bmatrix} 0 & -A_z \\ -A_z & 0 \end{bmatrix}
$$

(14.25)

These other components of the potential are unaffected by the 2-dimensional rotation because that 2-dimensional rotation happens in a 2-dimensional space that is distinct from and separate from the 4-dimensional space-time in which we sit. Only rotation in a separate space would leave the other components of the potential unaffected.

When we differentiate the 4-potential, we differentiate with respect to $[\partial t, \partial x, \partial y, \partial z]$ and not with respect to $[\partial t, -\partial x, -\partial y, -\partial z]$. The Lorentz transformation of these for a boost in the x-direction is:

$$\begin{bmatrix} \gamma_x & v_x\gamma_x \\ v_x\gamma_x & \gamma_x \end{bmatrix}\begin{bmatrix} \partial t & \partial x \\ \partial x & \partial t \end{bmatrix} = \begin{bmatrix} \gamma_x\partial t + v_x\gamma_x\partial x & \gamma_x\partial x + v_x\gamma_x\partial t \\ \gamma_x\partial x + v_x\gamma_x\partial t & \gamma_x\partial t + v_x\gamma_x\partial x \end{bmatrix} \quad (14.26)$$

Using the 'six separate 2-dimensional spaces' assumption, we have:

$$\begin{bmatrix} 0 & \partial y \\ \partial y & 0 \end{bmatrix}' = \begin{bmatrix} 0 & \partial y \\ \partial y & 0 \end{bmatrix}$$

$$\begin{bmatrix} 0 & \partial z \\ \partial z & 0 \end{bmatrix}' = \begin{bmatrix} 0 & \partial z \\ \partial z & 0 \end{bmatrix} \quad (14.27)$$

We see these as differential operators. We now form the transformed electromagnetic tensor as:

$$F_{\mu\nu}' = \frac{\partial \Phi_\nu'}{\partial x_\mu'} - \frac{\partial \Phi_\mu'}{\partial x_\nu'} \quad (14.28)$$

Wherein we use the rotated (boosted) components of the potential and the rotated (boosted) differentials. To do this, we need to separate the t-variables from the x-variables within the rotated matrices. This is:

$$\begin{bmatrix} \phi & 0 \\ 0 & \phi \end{bmatrix}' = \begin{bmatrix} \gamma_x\phi - v_x\gamma_x A_x & 0 \\ 0 & \gamma_x\phi - v_x\gamma_x A_x \end{bmatrix}$$

$$\begin{bmatrix} 0 & -A_x \\ -A_x & 0 \end{bmatrix}' = \begin{bmatrix} 0 & -\gamma_x A_x + v_x\gamma_x\phi \\ -\gamma_x A_x + v_x\gamma_x\phi & 0 \end{bmatrix} \quad (14.29)$$

And:

$$\begin{bmatrix} \partial t & 0 \\ 0 & \partial t \end{bmatrix}' = \begin{bmatrix} \gamma_x\partial t + v_x\gamma_x\partial x & 0 \\ 0 & \gamma_x\partial t + v_x\gamma_x\partial x \end{bmatrix}$$

$$\begin{bmatrix} 0 & \partial x \\ \partial x & 0 \end{bmatrix}' = \begin{bmatrix} 0 & \gamma_x\partial x + v_x\gamma_x\partial t \\ \gamma_x\partial x + v_x\gamma_x\partial t & 0 \end{bmatrix} \quad (14.30)$$

This separation is essential to the conventional presentation, but it seems arbitrary.

Aside: If we try to avoid this arbitrary separation by rotating the components separately, we would get:

$$\begin{bmatrix} \phi & 0 \\ 0 & \phi \end{bmatrix}' = \begin{bmatrix} \gamma_x & v_x\gamma_x \\ v_x\gamma_x & \gamma_x \end{bmatrix}\begin{bmatrix} \phi & 0 \\ 0 & \phi \end{bmatrix}$$
$$= \begin{bmatrix} \gamma_x\phi & v_x\gamma_x\phi \\ v_x\gamma_x\phi & \gamma_x\phi_x \end{bmatrix} \tag{14.31}$$

$$\begin{bmatrix} 0 & -A_x \\ -A_x & 0 \end{bmatrix}' = \begin{bmatrix} -v_x\gamma_x A_x & -\gamma_x A_x \\ -\gamma_x A_x & -v_x\gamma_x A_x \end{bmatrix} \tag{14.32}$$

$$\begin{bmatrix} \gamma_x & v_x\gamma_x \\ v_x\gamma_x & \gamma_x \end{bmatrix}\begin{bmatrix} \partial t & 0 \\ 0 & \partial t \end{bmatrix} = \begin{bmatrix} \gamma_x\partial t & v_x\gamma_x\partial t \\ v_x\gamma_x\partial t & \gamma_x\partial t \end{bmatrix} \tag{14.33}$$

$$\begin{bmatrix} \gamma_x & v_x\gamma_x \\ v_x\gamma_x & \gamma_x \end{bmatrix}\begin{bmatrix} 0 & \partial x \\ \partial x & 0 \end{bmatrix} = \begin{bmatrix} v_x\gamma_x\partial x & \gamma_x\partial x \\ \gamma_x\partial x & v_x\gamma_x\partial x \end{bmatrix} \tag{14.34}$$

These do not give the observed Lorentz transformations.

Using the transformed differentials (14.27) & (14.30) and the transformed potentials (14.25) & (14.29) the calculation of the Lorentz transformations follows; for example:

$$F_{12}' = \gamma_x^2 \left(\begin{bmatrix} \partial t + v_x\partial x & 0 \\ 0 & \partial t + v_x\partial x \end{bmatrix}\begin{bmatrix} 0 & -A_x + v_x\phi \\ -A_x + v_x\phi & 0 \end{bmatrix} - \begin{bmatrix} 0 & \partial x + v_x\partial t \\ \partial x + v_x\partial t & 0 \end{bmatrix}\begin{bmatrix} \phi - v_x A_x & 0 \\ 0 & \phi - v_x A_x \end{bmatrix} \right) \tag{14.35}$$

$$= \gamma_x^{\,2} \left(\begin{bmatrix} 0 & -\dfrac{\partial A_x}{\partial t} + v_x \dfrac{\partial \phi}{\partial t} - v_x \dfrac{\partial A_x}{\partial x} + v_x^{\,2} \dfrac{\partial \phi}{\partial x} \\ \sim & 0 \end{bmatrix} - \begin{bmatrix} 0 & \dfrac{\partial \phi}{\partial x} - v_x \dfrac{\partial A_x}{\partial x} + v_x \dfrac{\partial \phi}{\partial t} - v_x^{\,2} \dfrac{\partial A_x}{\partial t} \\ \sim & 0 \end{bmatrix} \right)$$

$$= \gamma_x^{\,2} \begin{bmatrix} 0 & \left(v_x^{\,2}-1\right)\dfrac{\partial \phi}{\partial x} + \left(v_x^{\,2}-1\right)\dfrac{\partial A_x}{\partial t} \\ \sim & 0 \end{bmatrix} \tag{14.36}$$

$$= \begin{bmatrix} 0 & -\dfrac{\partial \phi}{\partial x} - \dfrac{\partial A_x}{\partial t} \\ -\dfrac{\partial \phi}{\partial x} - \dfrac{\partial A_x}{\partial t} & 0 \end{bmatrix}$$

Wherein we have squiggled (~) out the duplicate information for presentational ease. Comparing this with the above untransformed electromagnetic tensor, (14.7), shows:

$$F_{12}{}' = F_{12} \tag{14.37}$$

Another example is:

$$F_{13}{}' = \gamma_x \left(\begin{bmatrix} \partial t + v_x \partial x & 0 \\ 0 & \partial t + v_x \partial x \\ -\begin{bmatrix} 0 & \partial y \\ \partial y & 0 \end{bmatrix} \end{bmatrix} \begin{bmatrix} 0 & -A_y \\ -A_y & 0 \\ \phi - v_x A_x & 0 \\ 0 & \phi - v_x A_x \end{bmatrix} \right) \tag{14.38}$$

$$= \gamma_x \left(\begin{bmatrix} 0 & -\dfrac{\partial A_y}{\partial t} - v_x \dfrac{\partial A_y}{\partial x} \\ \sim & 0 \end{bmatrix} - \begin{bmatrix} 0 & \dfrac{\partial \phi}{\partial y} - v_x \dfrac{\partial A_x}{\partial y} \\ \sim & 0 \end{bmatrix} \right)$$

$$= \gamma_x \begin{bmatrix} 0 & -\dfrac{\partial A_y}{\partial t} - \dfrac{\partial \phi}{\partial y} + v_x \left(\dfrac{\partial A_x}{\partial y} - \dfrac{\partial A_y}{\partial x} \right) \\ \sim & 0 \end{bmatrix} \tag{14.39}$$

Comparing this to (14.7) gives:

$$F_{13}' = \gamma\left(F_{13} + v_x F_{23}\right) \qquad (14.40)$$

We get the transformations of the fields as:

$$E_x' = E_x \quad : \quad B_x' = B_x$$
$$E_y' = \gamma\left(E_y - v_x B_z\right) \quad : \quad B_y' = \gamma\left(B_y + v_x E_z\right) \qquad (14.41)$$
$$E_z' = \gamma\left(E_z + v_x B_y\right) \quad : \quad B_z' = \gamma\left(B_z - v_x E_y\right)$$

Which concurs with the standard result. Since we have done this the standard way, we ought not to be surprised that it concurs with the standard result.

The quaternion version of Lorentz transformations:

The reader might think that since we calculated the electromagnetic fields from a quaternion potential using a 4-dimensional curl, we should do something similar to get the quaternion version of the Lorentz transformations. To do this, we would have to act upon the quaternion potential with a 4-dimensional space-time rotation matrix. We have seen that such rotations affect the components of the potential in directions outside of the plane of the rotation. We have seen that we do not have a 4×4 rotation matrix within the space-time in which we sit although we have them within the A_3 spaces which we think sum to form the space-time in which we sit. The simple fact is that the A_3 rotation matrices are not within the same algebra as the quaternions, although they are very similar. To get the right Lorentz transformations of the fields, we need to make the assumption of six separate 2-dimensional spaces in the space-time in which we sit. We need to make this assumption in the conventional presentation, and we need to make this assumption in the quaternion presentation.

It seems to your author that this need to use 2-dimensional rotation matrices to get the observed Lorentz force laws and the impossibility of using 4-dimensional rotation matrices indicates the presence of independent 2-dimensional spaces within our 4-dimensional space-time.

The Lorentz force laws:

Having obtained the Lorentz transformations of the fields, either conventionally or by some other means, the Lorentz force laws follow automatically from the assumption of the Coulomb force law. The Coulomb force law is:

$$\vec{F} = q\vec{E} \tag{14.42}$$

The electric charge, q, is invariant under a Lorentz transformation (one wonders why). In the case of E_y boosted in the (t, x) plane, we have:

$$F_y = q\gamma_x \left(E_y - v_x B_z \right) \tag{14.43}$$

For low velocities, $\gamma_x \approx 1$ and similarly to the above, the observed force laws follow:

$$F_x = q\left(E_x + v_y B_z - v_z B_y \right)$$
$$F_y = q\left(E_y + v_z B_x - v_x B_z \right) \tag{14.44}$$
$$F_z = q\left(E_z + v_x B_y - v_y B_x \right)$$

Fields and force in the time direction:

Neither the conventional presentation of electromagnetism nor the quaternion presentation of electromagnetism seems to make a lot of sense when we look at fields and forces in the time direction. From the Lorentz transformations, we have only three spatial forces and no temporal force in the time direction, F_t. Conventionally, based upon the electromagnetic 4-tensor, the Lorentz force law is in 4-vector form is written as:

$$F_\mu = qv^v F_{\mu v} \tag{14.45}$$

Which has the time component:

$$F_t = q\left(-v_x E_x - v_y E_y - v_z E_z\right) \tag{14.46}$$

Since we travel through time at velocity c, the force accelerating us in the time direction is zero. (The mantra is that the 4-accelleration is perpendicular to the 4-velocity. The mantra is correct in essence, but 4-vectors are not proper mathematical objects.) So, from this observation, we set:

$$F_t = 0$$
$$v_x E_x + v_y E_y + v_z E_z = 0 \tag{14.47}$$

However, this time component is added in because it seems to fit not because it follows from the Lorentz transformations. If we take the E_t field from the conjugate quaternion potential:

$$E_t = \frac{\partial \phi}{\partial t} - \frac{\partial A_x}{\partial x} - \frac{\partial A_y}{\partial y} - \frac{\partial A_z}{\partial z} \tag{14.48}$$

And we transform it as we did the other components of the $B \& E$ fields, we get:

$$E_t' = \gamma_x^2 \left(\begin{bmatrix} \partial t + v_x \partial x & 0 \\ 0 & \partial t + v_x \partial x \end{bmatrix} \begin{bmatrix} \phi - v_x A_x & 0 \\ 0 & \phi - v_x A_x \end{bmatrix} \right.$$
$$\left. - \begin{bmatrix} 0 & \partial x + v_x \partial t \\ \partial x + v_x \partial t & 0 \end{bmatrix} \begin{bmatrix} 0 & -A_x + v_x \phi \\ -A_x + v_x \phi & 0 \end{bmatrix} \right)$$
$$- \begin{bmatrix} 0 & \partial y \\ \partial y & 0 \end{bmatrix} \begin{bmatrix} 0 & -A_y \\ -A_y & 0 \end{bmatrix} - \begin{bmatrix} 0 & \partial z \\ \partial z & 0 \end{bmatrix} \begin{bmatrix} 0 & -A_z \\ -A_z & 0 \end{bmatrix} \tag{14.49}$$

$$= \gamma_x^2 \left(\left[\begin{array}{cc} \dfrac{\partial \phi}{\partial t}\left(1-v_x^2\right) + \dfrac{\partial A_x}{\partial x}\left(1-v_x^2\right) & 0 \\ 0 & \sim \end{array} \right] \right) - \left[\begin{array}{cc} -\dfrac{\partial A_y}{\partial y} - \dfrac{\partial A_z}{\partial z} & 0 \\ 0 & -\dfrac{\partial A_y}{\partial y} - \dfrac{\partial A_z}{\partial z} \end{array} \right]$$

(14.50)

$$= \dfrac{\partial \phi}{\partial t} + \dfrac{\partial A_x}{\partial x} + \dfrac{\partial A_y}{\partial y} + \dfrac{\partial A_z}{\partial z} \qquad (14.51)$$

The change of signs is quite surprising, but I think I've done the calculation correctly. We can multiply this transformed E_t by q and assume this is the force in the time direction, which we know is zero. This leads to:

$$\dfrac{\partial \phi}{\partial t} + \nabla \bullet \vec{A} = 0 \qquad (14.52)$$

Which is the Lorentz condition. Hm!, your author is unsure of these time component bits of this presentation.

Inhomo. Maxwell eqns. and the continuity eqn.:
Conventionally, the inhomogeneous Maxwell equations are assumed to be true because this fits with observation. The assumption is a little obscure because they are not assumed directly; instead, we assume a Lagrangian of the form:

$$\mathcal{L} = -\dfrac{1}{4} F_{\mu\nu} F^{\mu\nu} - J^\mu A_\mu \qquad (14.53)$$

and we derive the inhomogeneous Maxwell equations from this by varying the potential while leaving the current density constant. Conventionally, the continuity equation is a statement that electric charge is conserved. Along with the inhomogeneous Maxwell equations, it can be derived from the assumed Lagrangian (14.53), and so we see that the one assumption of the above Lagrangian leads to both the continuity equation and the inhomogeneous Maxwell equations.

The continuity equation is:

$$\frac{\partial \rho}{\partial t} + \frac{\partial J_x}{\partial x} + \frac{\partial J_y}{\partial y} + \frac{\partial J_z}{\partial z} = 0 \qquad (14.54)$$

The inhomogeneous Maxwell equations are (with physical constants set to unity):

$$\frac{\partial E_x}{\partial x} + \frac{\partial E_y}{\partial y} + \frac{\partial E_z}{\partial z} = \rho$$

$$\frac{\partial B_z}{\partial y} - \frac{\partial B_y}{\partial z} - \frac{\partial E_x}{\partial t} = J_x$$

$$\frac{\partial B_x}{\partial z} - \frac{\partial B_z}{\partial x} - \frac{\partial E_y}{\partial t} = J_y \qquad (14.55)$$

$$\frac{\partial B_y}{\partial x} - \frac{\partial B_x}{\partial y} - \frac{\partial E_z}{\partial t} = J_z$$

Putting these into the continuity equation gives:

$$\frac{\partial^2 E_x}{\partial t \partial x} - \frac{\partial^2 E_x}{\partial t \partial x} + \frac{\partial^2 E_y}{\partial t \partial y} - \frac{\partial^2 E_y}{\partial t \partial y} + \frac{\partial^2 E_z}{\partial t \partial z} - \frac{\partial^2 E_z}{\partial t \partial z}$$

$$+ \frac{\partial^2 B_z}{\partial x \partial y} - \frac{\partial^2 B_z}{\partial x \partial y} + \frac{\partial^2 B_y}{\partial x \partial z} - \frac{\partial^2 B_y}{\partial x \partial z} + \frac{\partial^2 B_x}{\partial y \partial z} - \frac{\partial^2 B_x}{\partial y \partial z} = 0 \qquad (14.56)$$

Which is obviously true. While ever we define the current densities as above (14.55) by the inhomogeneous Maxwell equations, we will automatically get the continuity equation – conservation of electric charge – and we do not need to define the continuity equation separately. Defining the continuity equation may lead to the inhomogeneous Maxwell equations, but it need not necessarily lead to them; other combinations of the derivatives of electric and magnetic fields would also satisfy the continuity equation.

In both the conventional presentation and the quaternion presentation, we have been unable to deduce the inhomogeneous Maxwell equations. In the conventional case, we have to assume a particular Lagrangian; in the quaternion case, we have to assume the inhomogeneous Maxwell equations. Both ways are just assuming the inhomogeneous Maxwell equations. In the next chapter, we will 'deduce' the inhomogeneous Maxwell equations as expectation equations.

Anti-matter conventionally:

There is an assumed 4-potential, $\Phi = \begin{bmatrix} \phi & -A_x & -A_y & -A_z \end{bmatrix}$ which leads to the conventional electromagnetic fields:

$$\Phi = \begin{bmatrix} \phi & -A_x & -A_y & -A_z \end{bmatrix}$$

$$
\begin{aligned}
E_x &= -\frac{\partial A_x}{\partial t} - \frac{\partial \phi}{\partial x} & B_x &= \left(\frac{\partial A_z}{\partial y} - \frac{\partial A_y}{\partial z} \right) \\
E_y &= -\frac{\partial A_y}{\partial t} - \frac{\partial \phi}{\partial y} & \quad : \quad & B_y &= \left(\frac{\partial A_x}{\partial z} - \frac{\partial A_z}{\partial x} \right) \\
E_z &= -\frac{\partial A_z}{\partial t} - \frac{\partial \phi}{\partial z} & B_z &= \left(\frac{\partial A_y}{\partial x} - \frac{\partial A_x}{\partial y} \right)
\end{aligned}
$$

(14.57)

If we reverse the signs of this potential, we get:

$$\Phi = \begin{bmatrix} -\phi & A_x & A_y & A_z \end{bmatrix}$$

$$
\begin{aligned}
E_x &= \frac{\partial A_x}{\partial t} + \frac{\partial \phi}{\partial x} & B_x &= -\left(\frac{\partial A_z}{\partial y} - \frac{\partial A_y}{\partial z} \right) \\
E_y &= \frac{\partial A_y}{\partial t} + \frac{\partial \phi}{\partial y} & \quad : \quad & B_y &= -\left(\frac{\partial A_x}{\partial z} - \frac{\partial A_z}{\partial x} \right) \\
E_z &= \frac{\partial A_z}{\partial t} + \frac{\partial \phi}{\partial z} & B_z &= -\left(\frac{\partial A_y}{\partial x} - \frac{\partial A_x}{\partial y} \right)
\end{aligned}
$$

(14.58)

We see that the electric field is reversed, as is the magnetic field. This shows the, quite obvious, fact that an anti-matter potential produces an anti-matter electric field but that the anti-matter magnetic field is no different from that which could be produced by reversing the 'normal matter' electric current. However, it would be prettier if we could derive the existence of anti-matter without the arbitrary reversal of the potential. We will look at a possible genesis of anti-matter in the next section when we consider the anti-quaternions.

Other 'sensible' potentials are:

$$\Phi = \begin{bmatrix} \phi & A_x & A_y & A_z \end{bmatrix}$$

$$E_x = \frac{\partial A_x}{\partial t} - \frac{\partial \phi}{\partial x} \qquad B_x = -\left(\frac{\partial A_z}{\partial y} - \frac{\partial A_y}{\partial z} \right)$$

$$E_y = \frac{\partial A_y}{\partial t} - \frac{\partial \phi}{\partial y} \qquad : \qquad B_y = -\left(\frac{\partial A_x}{\partial z} - \frac{\partial A_z}{\partial x} \right) \qquad (14.59)$$

$$E_z = \frac{\partial A_z}{\partial t} - \frac{\partial \phi}{\partial z} \qquad B_z = -\left(\frac{\partial A_y}{\partial x} - \frac{\partial A_x}{\partial y} \right)$$

And:

$$\Phi = \begin{bmatrix} -\phi & -A_x & -A_y & -A_z \end{bmatrix}$$

$$E_x = -\frac{\partial A_x}{\partial t} + \frac{\partial \phi}{\partial x} \qquad B_x = \left(\frac{\partial A_z}{\partial y} - \frac{\partial A_y}{\partial z} \right)$$

$$E_y = -\frac{\partial A_y}{\partial t} + \frac{\partial \phi}{\partial y} \qquad : \qquad B_y = \left(\frac{\partial A_x}{\partial z} - \frac{\partial A_z}{\partial x} \right) \qquad (14.60)$$

$$E_z = -\frac{\partial A_z}{\partial t} + \frac{\partial \phi}{\partial z} \qquad B_z = \left(\frac{\partial A_y}{\partial x} - \frac{\partial A_x}{\partial y} \right)$$

We see that we have four different definitions of the electric field but only two definitions of magnetic field. All of these lead to a consistent presentation of electromagnetism either conventionally or through quaternions, and all lead to the standard homogeneous Maxwell equations as a differential identity.

Anti-quaternions and anti-matter:

There are two quaternion algebras within the $C_2 \times C_2$ group. One of these, the quaternions, we have considered above. We call the other quaternion algebra the anti-quaternions. The commutation relations of the anti-quaternions are the reverse of the $SU(2)$ relations of the quaternions, and so we have an 'anti-$SU(2)$' Lie group. We can do with the anti-quaternions exactly what we did with the quaternions and derive the anti-quaternion $B \& E$ fields. Adopting the conjugate anti-quaternion potential to match what we did above with the quaternions and differentiating with respect to an anti-quaternion leads to:

$$E_t = \frac{\partial \phi}{\partial t} - \frac{\partial A_x}{\partial x} - \frac{\partial A_y}{\partial y} - \frac{\partial A_z}{\partial z} \qquad\qquad B_t = 0$$

$$E_x = -\frac{\partial A_x}{\partial t} - \frac{\partial \phi}{\partial x} \qquad\qquad B_x = -\left(\frac{\partial A_z}{\partial y} - \frac{\partial A_y}{\partial z} \right)$$

$$E_y = -\frac{\partial A_y}{\partial t} - \frac{\partial \phi}{\partial y} \qquad\qquad B_y = -\left(\frac{\partial A_x}{\partial z} - \frac{\partial A_z}{\partial x} \right) \qquad (14.61)$$

$$E_z = -\frac{\partial A_z}{\partial t} - \frac{\partial \phi}{\partial z} \qquad\qquad B_z = -\left(\frac{\partial A_y}{\partial x} - \frac{\partial A_x}{\partial y} \right)$$

We have the same electric field as we derived from the conjugate quaternion potential, but the magnetic field is reversed.

Imagine an electron moving across a magnetic field in such a way that it is deflected to the left. If the electron had a reversed magnetic moment, it would be deflected to the right. We posit that an electron in quaternion space is an electron but that an electron in anti-quaternion space, though still an electron, appears to observers as a positron. We have the prediction of anti-matter from the $C_2 \times C_2$ group. We are driven to posit that the existence of the anti-quaternions is the reason for the existence of anti-matter.

The reader is reminded that the quaternion algebra differs from the anti-quaternion algebra by the sign of one scaling parameter which we associated above, (8.18), with the charge of the electron.

Gauge invariance:
The electric and magnetic fields derived from the 4-potential in the form of the electromagnetic 4-tensor are gauge invariant. This means that we can add a gradient to the potential without affecting the derived fields. We have that the 4-tensor of the 4-potential:

$$\left[\phi + \frac{\partial \lambda}{\partial t}, -A_x - \frac{\partial \lambda}{\partial x}, -A_y - \frac{\partial \lambda}{\partial y}, -A_z - \frac{\partial \lambda}{\partial z} \right] \quad : \quad \lambda(x,y,z) \quad (14.62)$$

is the same as the 4-tensor (14.7) of the 4-potential (14.1). Note that this is not the case if λ is a function of time. If we allow λ to be a function of time as well as a function of the spatial co-ordinates, we get the fields as:

$$E_x = -\frac{\partial A_x}{\partial t} - \frac{\partial \phi}{\partial x} - 2\frac{\partial^2 \lambda}{\partial t \partial x} \qquad B_x = \left(\frac{\partial A_z}{\partial y} - \frac{\partial A_y}{\partial z} \right)$$

$$E_y = -\frac{\partial A_y}{\partial t} - \frac{\partial \phi}{\partial y} - 2\frac{\partial^2 \lambda}{\partial t \partial y} \qquad B_y = \left(\frac{\partial A_x}{\partial z} - \frac{\partial A_z}{\partial x} \right) \qquad (14.63)$$

$$E_z = -\frac{\partial A_z}{\partial t} - \frac{\partial \phi}{\partial z} - 2\frac{\partial^2 \lambda}{\partial t \partial z} \qquad B_z = \left(\frac{\partial A_y}{\partial x} - \frac{\partial A_x}{\partial y} \right)$$

Similarly adding a gradient to a quaternion potential and deriving the fields through non-commutative differentiation gives invariant B fields and the E fields:

$$E_t = \frac{\partial \phi}{\partial t} - \frac{\partial A_x}{\partial x} - \frac{\partial A_y}{\partial y} - \frac{\partial A_z}{\partial z} + \left(\frac{\partial^2 \lambda}{\partial t^2} - \frac{\partial^2 \lambda}{\partial x^2} - \frac{\partial^2 \lambda}{\partial y^2} - \frac{\partial^2 \lambda}{\partial z^2} \right) \quad (14.64)$$

$$E_x = -\frac{\partial \phi}{\partial x} - \frac{\partial A_x}{\partial t} - 2\frac{\partial^2 \lambda}{\partial t \partial x}$$

$$E_y = -\frac{\partial \phi}{\partial y} - \frac{\partial A_y}{\partial t} - 2\frac{\partial^2 \lambda}{\partial t \partial y} \qquad (14.65)$$

$$E_z = -\frac{\partial \phi}{\partial z} - \frac{\partial A_z}{\partial t} - 2\frac{\partial^2 \lambda}{\partial t \partial z}$$

We have the time component that we do not get by conventional means. We see that, if λ is not a function of time, the spatial components of the E field are invariant, as with the conventional presentation, but that the time component of the E field is not invariant – worth a Hm!

Summary:

The equations of electromagnetism are conventionally derived as the equations of motion of an assumed Lagrangian. Having calculated these equations of motion, we conventionally use a 12lb hammer to force the $SU(2)$ commutation relations on to the equations of motion. 12lb hammers have no place in theoretical physics. (They are very useful to engineers who build particle colliders.) The quaternion presentation has the $SU(2)$ commutation relations intrinsically within electromagnetism; we do not need the 12lb hammer. The quaternion presentation brings with it a 4-dimensional differential operator in place of the traditional curl that is conventionally used. The 4-dimensional differential operator produces all eight components of the electromagnetic field whereas the traditional curl produces only six of these eight. The quaternion presentation gives the Maxwell equations as a single quaternion equation rather than the conventional vector equation and scalar equation. The quaternion presentation automatically produces anti-matter. The quaternion presentation presents the electric field and the magnetic field as two separate, but connected, vectors. This contrasts with the 2nd rank 4-tensor of the conventional presentation. There is no such thing as a second rank tensor within a single division algebra, but there is such a thing as two connected vectors; however, in the next chapter, we do find a second rank tensor when the two quaternion algebras are aggregated together.

The two quaternion algebras derive from the finite group $C_2 \times C_2$. Also from that group derive the six A_3 algebras. These algebras provide the only explanation so far given for the nature of the distance function of the space-time in which we sit, and so, if this explanation is valid, we have unification between electromagnetism and the nature of space. We will later see that the A_3 spaces can be used to present general relativity.

Chapter 15

Tensors from $C_2 \times C_2$

Within this chapter, remarkable things happen. We do not really understand why these things are happening, but it seems that there is a relationship between an algebra and its anti-algebra that is connected to tensors. This allows us to derive the electromagnetic tensor. Of course, the anti-algebra of an algebra is just the algebra written in a different basis.

Direct derivation of the electromagnetic tensor:

There is an intriguing property of the electromagnetic field tensor; it looks like it aspires to be a quaternion[59]. If we ignore the difference between the electric field and the magnetic field[60], and we look at only the directions of the fields, we have a quaternion with zero real part.

$$F_{\mu\nu} = \begin{bmatrix} 0 & E_x & E_y & E_z \\ -E_x & 0 & -B_z & B_y \\ -E_y & B_z & 0 & -B_x \\ -E_z & -B_y & B_x & 0 \end{bmatrix} \sim \begin{bmatrix} 0 & x & y & z \\ -x & 0 & -z & y \\ -y & z & 0 & -x \\ -z & -y & x & 0 \end{bmatrix} \quad (15.1)$$

If we had defined the magnetic field to point in the opposite direction by defining the curl oppositely or the B field oppositely or whatever, we would have the modified electromagnetic tensor as:

---

[59] Oh! such unbridled optimism and ambition!
[60] One observer's magnetic field is another observer's electric field.

$$F_{\mu\nu}^{\,Modified} = \begin{bmatrix} 0 & E_x & E_y & E_z \\ -E_x & 0 & B_z & -B_y \\ -E_y & -B_z & 0 & B_x \\ -E_z & B_y & -B_x & 0 \end{bmatrix} \sim \begin{bmatrix} 0 & x & y & z \\ -x & 0 & z & -y \\ -y & -z & 0 & x \\ -z & y & -x & 0 \end{bmatrix} \quad (15.2)$$

This modified electromagnetic tensor aspires to be an anti-quaternion. To concur with convention, we will ignore the modified version of the electromagnetic tensor and stick with the conventional direction of the magnetic field.

Viewed as a matrix, the electromagnetic tensor apparently has commutation relations:

$$\left[E_x, E_y\right] = B_z, \quad \left[E_x, E_z\right] = -B_y, \quad \left[E_y, E_z\right] = B_x$$
$$\left[B_x, B_y\right] = B_z, \quad \left[B_x, B_z\right] = -B_y, \quad \left[B_y, B_z\right] = B_x$$

$$\left[E_x, B_x\right] = 0, \quad \left[E_x, B_y\right] = E_z, \quad \left[E_x, B_z\right] = -E_y$$
$$\left[E_y, B_x\right] = -E_z, \quad \left[E_y, B_y\right] = 0, \quad \left[E_y, B_z\right] = E_x$$
$$\left[E_z, B_x\right] = E_y, \quad \left[E_z, B_y\right] = -E_x, \quad \left[E_z, B_z\right] = 0$$

(15.3)

However, the electromagnetic tensor is not a division algebra. A pedant would insist that multiplication is meaningless outside of a group or a division algebra[61], and so these commutation relations are not commutation relations. They are certainly not an established Lie algebra. There are also too many of them, and the electric field components are not closed; the product of two electric field components is not an electric field component. We would like to ignore the commutation relations above, (15.3), and take the view that the electromagnetic tensor is not quantitised. This fits with the conventional view of the electromagnetic tensor.

When we rewrote electromagnetism, we derived the electric field and the magnetic field from a conjugate quaternion potential as two

---

[61] Mating a cow with a donkey to produce a duck is not *bona fide* multiplication.

separate 4×4 quaternion matrices. As such, we derived four components to each field. We also derived the E field and the B field of the anti-quaternions from a conjugate anti-quaternion potential as two separate 4×4 anti-quaternion matrices. We used the conjugate quaternion potential to fit with the conventional definition of the electric and magnetic field, and we used the conjugate anti-quaternion potential to be consistent. We continue to use the conjugate potentials to fit with convention.

The E field of a conjugate quaternion potential is:

$$E_{Q^*} =$$

$$
\begin{bmatrix}
\left(\dfrac{\partial \phi}{\partial t} - \dfrac{\partial A_x}{\partial x} - \dfrac{\partial A_y}{\partial y} - \dfrac{\partial A_z}{\partial z} \right) & -\dfrac{\partial A_x}{\partial t} - \dfrac{\partial \phi}{\partial x} & -\dfrac{\partial A_y}{\partial t} - \dfrac{\partial \phi}{\partial y} & -\dfrac{\partial A_z}{\partial t} - \dfrac{\partial \phi}{\partial z} \\[2ex]
\dfrac{\partial A_x}{\partial t} + \dfrac{\partial \phi}{\partial x} & \sim & \dfrac{\partial A_z}{\partial t} + \dfrac{\partial \phi}{\partial z} & -\dfrac{\partial A_y}{\partial t} - \dfrac{\partial \phi}{\partial y} \\[2ex]
\dfrac{\partial A_y}{\partial t} + \dfrac{\partial \phi}{\partial y} & -\dfrac{\partial A_z}{\partial t} - \dfrac{\partial \phi}{\partial z} & \sim & \dfrac{\partial A_x}{\partial t} + \dfrac{\partial \phi}{\partial x} \\[2ex]
\dfrac{\partial A_z}{\partial t} + \dfrac{\partial \phi}{\partial z} & \dfrac{\partial A_y}{\partial t} + \dfrac{\partial \phi}{\partial y} & -\dfrac{\partial A_x}{\partial t} - \dfrac{\partial \phi}{\partial x} & \sim
\end{bmatrix}
\quad (15.4)
$$

The E field of a conjugate anti-quaternion potential is:

$$E_{Anti-Q^*} =$$

$$
\begin{bmatrix}
\left(\dfrac{\partial \phi}{\partial t} - \dfrac{\partial A_x}{\partial x} - \dfrac{\partial A_y}{\partial y} - \dfrac{\partial A_z}{\partial z} \right) & -\dfrac{\partial A_x}{\partial t} - \dfrac{\partial \phi}{\partial x} & -\dfrac{\partial A_y}{\partial t} - \dfrac{\partial \phi}{\partial y} & -\dfrac{\partial A_z}{\partial t} - \dfrac{\partial \phi}{\partial z} \\[2ex]
\dfrac{\partial A_x}{\partial t} + \dfrac{\partial \phi}{\partial x} & \sim & -\dfrac{\partial A_z}{\partial t} - \dfrac{\partial \phi}{\partial z} & \dfrac{\partial A_y}{\partial t} + \dfrac{\partial \phi}{\partial y} \\[2ex]
\dfrac{\partial A_y}{\partial t} + \dfrac{\partial \phi}{\partial y} & \dfrac{\partial A_z}{\partial t} + \dfrac{\partial \phi}{\partial z} & \sim & -\dfrac{\partial A_x}{\partial t} - \dfrac{\partial \phi}{\partial x} \\[2ex]
\dfrac{\partial A_z}{\partial t} + \dfrac{\partial \phi}{\partial z} & -\dfrac{\partial A_y}{\partial t} - \dfrac{\partial \phi}{\partial y} & \dfrac{\partial A_x}{\partial t} + \dfrac{\partial \phi}{\partial x} & \sim
\end{bmatrix}
\quad (15.5)
$$

Wherein we have squiggled out the duplicate information to ease presentation. These matrices are the quaternion algebra written in two different bases, and, as such, perhaps we ought not to add them together directly. However, we did such addition of different algebraic forms when we derived the distance function of the space-time in which we sit. If we do simply add these two E field matrices, because of the distribution of the minus signs within these matrices, we get:

$$E_{Q^*} + E_{Anti-Q^*} =$$

$$2 \begin{bmatrix} \left(\dfrac{\partial \phi}{\partial t} - \dfrac{\partial A_x}{\partial x} - \dfrac{\partial A_y}{\partial y} - \dfrac{\partial A_z}{\partial z} \right) & -\dfrac{\partial A_x}{\partial t} - \dfrac{\partial \phi}{\partial x} & -\dfrac{\partial A_y}{\partial t} - \dfrac{\partial \phi}{\partial y} & -\dfrac{\partial A_z}{\partial t} - \dfrac{\partial \phi}{\partial z} \\ \dfrac{\partial A_x}{\partial t} + \dfrac{\partial \phi}{\partial x} & \sim & 0 & 0 \\ \dfrac{\partial A_y}{\partial t} + \dfrac{\partial \phi}{\partial y} & 0 & \sim & 0 \\ \dfrac{\partial A_z}{\partial t} + \dfrac{\partial \phi}{\partial z} & 0 & 0 & \sim \end{bmatrix} \qquad (15.6)$$

Taking:

$$\frac{\partial \phi}{\partial t} - \frac{\partial A_x}{\partial x} - \frac{\partial A_y}{\partial y} - \frac{\partial A_z}{\partial z} = 0 \qquad (15.7)$$

we get the electric part of the electromagnetic field tensor:

$$E_{Q^*} + E_{Anti-Q^*} = 2 \begin{bmatrix} 0 & E_x & E_y & E_z \\ -E_x & 0 & 0 & 0 \\ -E_y & 0 & 0 & 0 \\ -E_z & 0 & 0 & 0 \end{bmatrix} \qquad (15.8)$$

Doing the same with the B field of the conjugate quaternion potential and the conjugate anti-quaternion potential gives:

$$B_{Q^*} + B_{Anti-Q^*} =$$

$$2\begin{bmatrix} 0 & 0 & 0 & 0 \\ 0 & 0 & \dfrac{\partial A_x}{\partial y} - \dfrac{\partial A_y}{\partial x} & \dfrac{\partial A_x}{\partial z} - \dfrac{\partial A_z}{\partial x} \\ 0 & -\left(\dfrac{\partial A_x}{\partial y} - \dfrac{\partial A_y}{\partial x}\right) & 0 & \dfrac{\partial A_y}{\partial z} - \dfrac{\partial A_z}{\partial y} \\ 0 & -\left(\dfrac{\partial A_x}{\partial z} - \dfrac{\partial A_z}{\partial x}\right) & -\left(\dfrac{\partial A_y}{\partial z} - \dfrac{\partial A_z}{\partial y}\right) & 0 \end{bmatrix} \quad (15.9)$$

$$B_{Q^*} + B_{Anti-Q^*} = 2\begin{bmatrix} 0 & 0 & 0 & 0 \\ 0 & 0 & -B_z & B_y \\ 0 & B_z & 0 & -B_x \\ 0 & -B_y & B_x & 0 \end{bmatrix} \quad (15.10)$$

Adding this sum of E fields and the sum of B fields gives the conventional electromagnetic field tensor:

$$E_{Q^*} + E_{Anti-Q^*} + B_{Q^*} + B_{Anti-Q^*} = 2\begin{bmatrix} 0 & E_x & E_y & E_z \\ -E_x & 0 & -B_z & B_y \\ -E_y & B_z & 0 & -B_x \\ -E_z & -B_y & B_x & 0 \end{bmatrix} \quad (15.11)$$

$$= 2\left(d_L^{Q^*} + d_L^{Anti-Q^*}\right)$$

We have a remarkable connection between the $C_2 \times C_2$ group and the electromagnetic tensor. We see here the emergence of classical physics from quantum physics. We might think that we have taken the average field and that the electromagnetic tensor is just the expectation 'value' of the quaternion $B \& E$ fields. Of course, no sensible mathematician would add two copies of an algebra written in two different bases; so why does it work? We don't know, but we think it might be the work of a 'clodhopping' macroscopic observer. We think that the classical universe emerges from the division algebras as an 'average' expectation universe.

We cannot justify the 'expectation universe' mathematically, but it does fit with observation.

We could have subtracted the E field of the anti-quaternion from the E field of the quaternion to give:

$$E_{Q^*} - E_{Anti-Q^*} = 2\begin{bmatrix} 0 & 0 & 0 & 0 \\ 0 & 0 & -E_z & E_y \\ 0 & E_z & 0 & -E_x \\ 0 & -E_y & E_x & 0 \end{bmatrix} \quad (15.12)$$

and the B field as:

$$B_{Q^*} - B_{Anti-Q^*} = 2\begin{bmatrix} 0 & B_x & B_y & B_z \\ -B_x & 0 & 0 & 0 \\ -B_y & 0 & 0 & 0 \\ -B_z & 0 & 0 & 0 \end{bmatrix} \quad (15.13)$$

Leading to:

$$E_{Q^*} - E_{Anti-Q^*} + B_{Q^*} - B_{Anti-Q^*} = 2\begin{bmatrix} 0 & B_x & B_y & B_z \\ -B_x & 0 & -E_z & E_y \\ -B_y & E_z & 0 & -E_x \\ -B_z & -E_y & E_x & 0 \end{bmatrix} \quad (15.14)$$

$$= 2\left(d_L^{Q^*} - d_L^{Anti-Q^*}\right)$$

Of course, subtracting the E field of the conjugate quaternion from the E field of the conjugate anti-quaternion and subtracting the B field of the conjugate quaternion from the B field of the conjugate anti-quaternion just reverses the signs in (15.14). We remind the reader that all of the above are based upon the conventional definition of the magnetic field.

We point out that the electromagnetic tensor is an anti-symmetric tensor with six independent components.

From microscopic physics to macroscopic physics:

The E field and the B field of a quaternion conjugate potential are both quantitised with the $SU(2)$ commutation relations; as are the anti-quaternion fields quantitised with the anti-$SU(2)$ commutation relations. If we hold the view that the electromagnetic tensor is not quantitised, then, by adding the E fields and the B fields as above, we have moved from quantitised electromagnetism to non-quantitised electromagnetism. A single electron will be in either the quaternion base or the anti-quaternion base and hence be quantitised. Of a million electrons, approximately half will be in each type of quaternion base, and so the aggregate will, we opine, be the sum of the different bases and not be quantitised. Perhaps this is the nature of the move from quantum mechanics to classical mechanics as we move from microscopic to macroscopic.

When we formed the electromagnetic tensor as the sum of the quaternion fields and the anti-quaternion field, we destroyed the algebraic structure of both the quaternions and the anti-quaternions. If there is no algebraic structure, there is no relationship between the different variables and it is meaningless to say that some variables are imaginary variables (square roots of plus unity or minus unity). We see this when we accept that, in spite of its appearance, the electromagnetic tensor is not a matrix. If the electromagnetic tensor were a matrix, then there would be a relationship between the elements of the opposing diagonal and the elements of the leading diagonal expressed as:

$$\begin{bmatrix} 0 & 0 & 0 & E_z \\ 0 & 0 & -B_z & 0 \\ 0 & B_z & 0 & 0 \\ -E_z & 0 & 0 & 0 \end{bmatrix}^2 = \begin{bmatrix} -E_z^2 & 0 & 0 & 0 \\ 0 & -B_z^2 & 0 & 0 \\ 0 & 0 & -B_z^2 & 0 \\ 0 & 0 & 0 & -E_z^2 \end{bmatrix} \quad (15.15)$$

and similarly with other elements. If the variables are not imaginary variables, then what are they? They have magnitude, but they are bereft of any relationship with other variables; they are real variables. It seems that, when we aggregate together division algebras in different bases and thereby destroy algebraic structure, we move from division algebra mathematics to the $\mathbb{R}^n$ type of mathematics. Four real variables are a manifold; there is no sense of parallel transport or of distance or of angle in a manifold; these properties have to be added to the manifold. So, it seems, the microscopic world is written in the mathematics of division algebras and the macroscopic world is written in the mathematics of $\mathbb{R}^n$. Of course, macroscopic (classical) physics has always been written in $\mathbb{R}^n$; in particular, general relativity is written in $\mathbb{R}^4$.

Looking at the last lines of (15.11) & (15.14), we seem to have a bias towards the left differentials. If we had chosen to define the magnetic field to point in the opposite direction to the conventional definition, we would have a similar bias towards the right differentials. However, we define the direction of the magnetic fields, we get a bias when we move from the individual algebras into the sum of the algebras.

Quantitising a classical field theory:
In quantum field theory, QFT, we begin with a classical field theory written with real variables and we quantitise this field by imposing commutation relations upon these real variables. Since a set of real variables does not have any commutation relations, we must be changing the nature of these variables by imposing commutation relations. It seems that, when we impose commutation relations upon a set of real variables, we are changing these real variables into imaginary variables – what else could they be? A pedantic mathematician would insist that multiplication does not exist outside of a division algebra (or group) and imaginary variables do not exist outside of a division algebra. So, being pedantic, when we quantitise a classical field theory, we really ought to specify to which division algebra the imaginary variables belong. If we quantitise the classical field theory by imposing the $SU(2)$ commutation relations, we are effectively specifying the quaternion division algebra as being the division algebra to which the imaginary variables belong. Of course,

in addition to the three non-commutative imaginary variables, the quaternion division algebra has the commutative real variable. When we have imposed commutation relations on the classical field theory, we have converted it into a quantum field theory. Within this quantum field theory, we indirectly promote the imaginary variables to operators which create or annihilate particles. Perhaps, we associate the quaternion imaginary variables with creation operators and the anti-quaternion imaginary variables with annihilation operators. Thus, with each imaginary variable, we associate a particular kind of particle. With $SU(2)$ we associate the $\{W^{\pm}, Z^{0}\}$ bosons. However, we have forgotten the commutative (real) variable of the quaternion algebra; surely, if the non-commutative variables are associated with their own particle, then the commutative variable should have its own particle. We might presume that particles associated with commutative variables are fermions and that the particle associated with the real quaternion variable is the electron, but now we are getting very speculative.

We see that whereas aggregation of division algebras moves from imaginary variables with commutation relations to real variables with no commutation relations, quantitisation of a classical field theory moves from real variables with no commutation relations to imaginary variables with commutation relations.

Equal scaling of quaternions and anti-quaternions:
We have glossed over something above. There is no reason to think that the quaternion algebra and the anti-quaternion algebra are the same size. When we added the $B \& E$ fields of these two algebras, we assumed that they were of the same size. This is the same as assuming the electric charge of a positron is equal, but opposite, to the electric charge of an electron; we assume the ε scaling parameter is of the same magnitude in both algebras. Because there is a definite quantum amount of minimal electric charge, the two algebras have to be the same size.

The Maxwell equations:

Within both the quaternions and the anti-quaternions, everything is a quaternion matrix or an anti-quaternion matrix respectively. In particular, second differentials like curly-B and straight-E are quaternion/anti-quaternion matrices. We have the basic differential identity $\{B,d\} = [E,d]$ which, within a particular algebra, is the homogeneous Maxwell equations. This differential identity is true in both the quaternions and the anti-quaternions. Of course, the B field in the two algebras is different by a minus sign; none-the-less, if we are allowed to add algebras in different bases (take the expectation algebra so to speak), it follows that:

$$\{B,d\}_Q + \{B,d\}_{Anti-Q} = [E,d]_Q + [E,d]_{Anti-Q} \qquad (15.16)$$

In terms of the electric and magnetic fields:

$$E_Q = \begin{bmatrix} E_t & E_x & E_y & E_z \\ -E_x & E_t & -E_z & E_y \\ -E_y & E_z & E_t & -E_x \\ -E_z & -E_y & E_x & E_t \end{bmatrix} \quad E_{Anti-Q} = \begin{bmatrix} E_t & E_x & E_y & E_z \\ -E_x & E_t & E_z & -E_y \\ -E_y & -E_z & E_t & E_x \\ -E_z & E_y & -E_x & E_t \end{bmatrix}$$

$$B_Q = \begin{bmatrix} 0 & B_x & B_y & B_z \\ -B_x & 0 & -B_z & B_y \\ -B_y & B_z & 0 & -B_x \\ -B_z & -B_y & B_x & 0 \end{bmatrix} \quad B_{Anti-Q} = \begin{bmatrix} 0 & B_x & B_y & B_z \\ -B_x & 0 & B_z & -B_y \\ -B_y & -B_z & 0 & B_x \\ -B_z & B_y & -B_x & 0 \end{bmatrix}$$

$$(15.17)$$

we have:

$$\{B,d\}_Q + \{B,d\}_{Anti-Q} = \begin{bmatrix} \dfrac{\partial B_x}{\partial x} + \dfrac{\partial B_y}{\partial y} + \dfrac{\partial B_y}{\partial y} & \dfrac{\partial B_x}{\partial t} & \dfrac{\partial B_y}{\partial t} & \dfrac{\partial B_z}{\partial t} \\ -\dfrac{\partial B_x}{\partial t} & \sim & 0 & 0 \\ -\dfrac{\partial B_y}{\partial t} & 0 & \sim & 0 \\ -\dfrac{\partial B_z}{\partial t} & 0 & 0 & \sim \end{bmatrix}$$

$$(15.18)$$

and:

$$[E,d]_Q + [E,d]_{Anti-Q} = \begin{bmatrix} 0 & 0 & 0 & 0 \\ 0 & 0 & \dfrac{\partial E_y}{\partial x} - \dfrac{\partial E_x}{\partial y} & \dfrac{\partial E_z}{\partial x} - \dfrac{\partial E_x}{\partial z} \\ 0 & -\left(\dfrac{\partial E_y}{\partial x} - \dfrac{\partial E_x}{\partial y}\right) & 0 & \dfrac{\partial E_z}{\partial y} - \dfrac{\partial E_y}{\partial z} \\ 0 & -\left(\dfrac{\partial E_z}{\partial x} - \dfrac{\partial E_x}{\partial z}\right) & -\left(\dfrac{\partial E_z}{\partial y} - \dfrac{\partial E_y}{\partial z}\right) & 0 \end{bmatrix}$$

$$(15.19)$$

At first appearances, these cannot be equal. However, if we combine these two matrices (15.18) & (15.19) together by addition to form a 'tensor' as we did to form the electromagnetic tensor, that 'tensor' aspires to be a quaternion. It can be a quaternion only if:

$$\frac{\partial B_x}{\partial t} = \frac{\partial E_y}{\partial z} - \frac{\partial E_z}{\partial y}$$

$$\frac{\partial B_y}{\partial t} = \frac{\partial E_z}{\partial x} - \frac{\partial E_x}{\partial z}$$

$$\frac{\partial B_z}{\partial t} = \frac{\partial E_x}{\partial y} - \frac{\partial E_y}{\partial x}$$

$$(15.20)$$

Which are the three conventional homogeneous Maxwell equations. We know these are true because they are no more than a differential identity, and so this 'making everything into a quaternion' seems compulsory.

Similarly with $\{E,d\} \& [B,d]$:

$$\{E,d\}_Q + \{E,d\}_{Anti-Q} =$$

$$\begin{bmatrix} \dfrac{\partial E_t}{\partial t} + \dfrac{\partial E_x}{\partial x} + \dfrac{\partial E_y}{\partial y} + \dfrac{\partial E_y}{\partial y} & \dfrac{\partial E_x}{\partial t} - \dfrac{\partial E_t}{\partial x} & \dfrac{\partial E_y}{\partial t} - \dfrac{\partial E_t}{\partial y} & \dfrac{\partial E_z}{\partial t} - \dfrac{\partial E_t}{\partial z} \\[2ex] -\left(\dfrac{\partial E_x}{\partial t} - \dfrac{\partial E_t}{\partial x} \right) & \sim & 0 & 0 \\[2ex] -\left(\dfrac{\partial E_y}{\partial t} - \dfrac{\partial E_t}{\partial y} \right) & 0 & \sim & 0 \\[2ex] -\left(\dfrac{\partial E_z}{\partial t} - \dfrac{\partial E_t}{\partial z} \right) & 0 & 0 & \sim \end{bmatrix} \quad (15.21)$$

And:

$$[B,d]_Q + [B,d]_{Anti-Q} =$$

$$\begin{bmatrix} 0 & 0 & 0 & 0 \\[2ex] 0 & 0 & \dfrac{\partial B_y}{\partial x} - \dfrac{\partial B_x}{\partial y} & \dfrac{\partial B_z}{\partial x} - \dfrac{\partial B_x}{\partial z} \\[2ex] 0 & -\left(\dfrac{\partial B_y}{\partial x} - \dfrac{\partial B_x}{\partial y} \right) & 0 & \dfrac{\partial B_z}{\partial y} - \dfrac{\partial B_y}{\partial z} \\[2ex] 0 & -\left(\dfrac{\partial B_z}{\partial x} - \dfrac{\partial B_x}{\partial z} \right) & -\left(\dfrac{\partial B_z}{\partial y} - \dfrac{\partial B_y}{\partial z} \right) & 0 \end{bmatrix} \quad (15.22)$$

The sum of these two matrices (15.21) & (15.22) is also a 'tensor', but it can be a quaternion only if we reverse the sign of the sum of the B fields. This is equivalent to putting:

$$\{E,d\}_Q + \{E,d\}_{Anti-Q} = -[B,d]_Q - [B,d]_{Anti-Q} \quad (15.23)$$

This leads to:

$$\frac{\partial E_x}{\partial t} - \frac{\partial E_t}{\partial x} = \frac{\partial B_z}{\partial y} - \frac{\partial B_y}{\partial z}$$

$$\frac{\partial E_y}{\partial t} - \frac{\partial E_t}{\partial y} = \frac{\partial B_x}{\partial z} - \frac{\partial B_z}{\partial x} \qquad (15.24)$$

$$\frac{\partial E_z}{\partial t} - \frac{\partial E_t}{\partial z} = \frac{\partial B_y}{\partial x} - \frac{\partial B_x}{\partial y}$$

These are the three conventional inhomogeneous Maxwell equations.

Have we deduced the inhomogeneous Maxwell equations, or have we fudged it? Both conventionally and quaternion-wise, we had, in the previous chapter, to assume the inhomogeneous Maxwell equations. We do not have to assume them now.

This seems to indicate that, while the homogeneous Maxwell equations are a basic truth at both quantum level and macroscopic level, the inhomogeneous Maxwell equations are a product of macroscopic observation aspiring to be a basic truth at quantum level.

Above, we have done more than take 'expectation tensors'. Above, we have, driven by the known truth of the homogeneous Maxwell equations, assumed the form of that 'expectation tensor'. It fits observation.

Philosophical mumblings:

Some parts of mathematics are discovered; examples are the real numbers, the complex numbers, $\mathbb{C}$, or the hyperbolic complex numbers, $\mathbb{S}$, which we found inside the group C_2. Of course, the group C_2 was discovered inside the real numbers; it is just the multiplicative relations between $\{-1, +1\}$. Other parts of mathematics are invented; examples are the various metric spaces defined by an invented distance function and mathematical constructions like the Clifford algebras. Some parts of mathematics seem to be invented but are later found to be discovered. An example of this are the Clifford algebras which are 'really' just $C_2 \times C_2 \times ...$

algebras in disguise. The quaternions, $\mathbb{H}$, are another example of this. They were invented by William Hamilton in the 1860's and were thought to be invented mathematical entities until we recently discovered them within the $C_2 \times C_2$ group. We now know the quaternions are discovered mathematical objects. Before we discovered that the electromagnetic tensor is just the sum of the E fields and B fields of the quaternions and the anti-quaternions, we saw tensors as invented mathematics; Riemann invented them. Of course, the quaternions were discovered within the $C_2 \times C_2$ group using 'proper mathematics' whereas the electromagnetic tensor was derived from the 'improper mathematics' of aggregating together algebras. Perhaps we will have to change our view that tensors are invented mathematics, or perhaps we will not have to change that view.

We can take the view that discovered mathematical objects 'really' exist whereas invented mathematical objects do not 'really' exist but are just the imaginings of wayward mathematicians. One might expect that the universe would be written in 'real' mathematics rather than in 'invented' mathematics. It is therefore remarkable that the 'invented' mathematics which we call tensors is so utterly central to physical theories like general relativity and electromagnetism. It might be that, as with the quaternions, tensors are discovered mathematics, or it might be that tensors are invented mathematics. It might be that, as seems to be the case with Lie groups, the idea is correct but a the basic assumption (the nature of space) is incorrect and so there is a confusion between invented and discovered mathematics.

Chapter 16

Anti-matter and the Unbalanced Universe

In the previous chapter, we saw that adding the $B \& E$ fields of the quaternions and the anti-quaternions produced a tensor which has the same distribution of minus signs as a quaternion:

$$E_{Q^*} + E_{Anti-Q^*} + B_{Q^*} + B_{Anti-Q^*}$$

$$= 2 \begin{bmatrix} 0 & E_x & E_y & E_z \\ -E_x & 0 & -B_z & B_y \\ -E_y & B_z & 0 & -B_x \\ -E_z & -B_y & B_x & 0 \end{bmatrix} \qquad (16.1)$$

Subtracting the anti-quaternion $B \& E$ fields from the quaternion $B \& E$ fields also produced a tensor which has the same distribution of minus signs as a quaternion, (15.14). If we had subtracted the quaternion fields from the anti-quaternion fields, we would have again produced a (conjugate) quaternion like tensor. However we do it, we get a quaternion like tensor and not an anti-quaternion like tensor. Similarly, when we aggregated together the fundamental differential identities of the quaternions and the anti-quaternions, (15.16), we found that the 'known truth' of the homogeneous Maxwell equations compelled us to the result being a quaternion and not an anti-quaternion. We associate the quaternions with matter and the anti-quaternions with anti-matter. We see that aggregation is 'unbalancing' the universe and preferring matter over anti-matter.

We have consistently, but arbitrarily, defined the B field as:

$$B = \frac{1}{2}(d_L - d_R) \qquad (16.2)$$

Looking at (15.11) and (15.14), we see that we had defined the B field oppositely as $d_R - d_L$, we would have been driven to an anti-quaternion rather than a quaternion. With such a definition of the B field, we would have found conventional electromagnetism (matter) to be within the anti-quaternions rather than within the quaternions. We see that whether we are driven to prefer a quaternion or an anti-quaternion when we aggregate the algebras depends upon the arbitrary choice of the direction of the B field, but that is not the important point.

The important point is that we are driven to prefer one algebra over the other by aggregation. Aggregation of the algebras has 'broken' the balance between the algebras. Aggregation has preferred matter over anti-matter. Moving from quantum mathematics to classical mathematics has 'unbalanced the mathematics.

We previously associated aggregation with the actions of a clod-hopping macroscopic observer. We are now faced with having to attribute the imbalance of matter and anti-matter with either the clod-hopping macroscopic observer or something else. Since it is unlikely that the clod-hopping macroscopic observer was around at the start of the universe, and we are not prepared to hypothesize the existence of a deity, we need something other than the clod-hopping macroscopic observer to explain why algebras aggregate together.

Why do algebras aggregate?

Perhaps aggregated algebras form a lower energy state (or a lower something state) than the separate algebras. We note that, considering the electromagnetic tensor as a matrix, the determinant of the electromagnetic tensor is:

$$Det_{(EMAG)} = \left(B_x E_x + B_y E_y + B_z E_z \right)^2 \qquad (16.3)$$

The determinant of the sum of the $B \& E$ fields is:

$$Det_{(E+B)} = \left(\begin{array}{c} E_x^2 + 2B_x E_x + B_x^2 + E_y^2 + 2B_y E_y \\ + B_y^2 + E_z^2 + 2B_z E_z + B_z^2 \end{array} \right)^2 \quad (16.4)$$

The determinant of the sum of the anti-quaternion $B \& E$ fields is similar to (16.4). Clearly, the determinant of the electromagnetic tensor is less than the determinants of the separate algebras, but we do not know what the determinant measures. Whatever the determinant measures, it is conserved under change of basis because the determinant in invariant under change of basis.

At this time, we do not know which aspects of the physical universe are 'classical' in that they derive from aggregation and which aspects of the physical universe are 'quantum' in that they are within the pertinent division algebras. It seems that the homogeneous Maxwell equations are both quantum and classical whereas the inhomogeneous Maxwell equations are only classical. It seems that the distance function of space-time $(+,-,-,-)$ is classical as is the complete Lorentz group $SO(3,1)$.

Is mass a classical thing that does not exist within the quantum universe? It seems more likely that classical mass is the momentum energy tensor, but that quantum mass is just a real number (the coefficient of a quadratic term in the Lagrangian). In electroweak theory, quantum mass is generated by the Higgs mechanism. The Higgs mechanism involves breaking a symmetry to make two complex numbers (a spinor) into four real numbers. Although a spinor is often presented in physics as a pair of complex numbers, a mathematical treatment of spinors (Clifford algebra) sees a spinor to be a quaternion. Under aggregation, a quaternion becomes four real numbers and the rotational symmetry of the quaternion, $SU(2)$, is broken. Perhaps this is the nature of the Higgs mechanism. Yes, there is much not yet understood in theoretical physics.

Chapter 17

In Quest of the Gravitation Tensor

Within general relativity, we have the Einstein tensor:

$$G^{\mu\nu} = R^{\mu\nu} - \frac{1}{2}g^{\mu\nu}R_S \qquad (17.1)$$

wherein $G^{\mu\nu}$ is the Einstein tensor, $R^{\mu\nu}$ is the Ricci tensor, $g^{\mu\nu}$ is the metric tensor and R_S is the Ricci scalar. All three of these tensors are symmetric with six independent elements. The field equations of general relativity are:

$$G^{\mu\nu} = 8\pi G T^{\mu\nu} \qquad (17.2)$$

wherein $T^{\mu\nu}$ is the symmetric mass-energy tensor and G is the gravitational constant. The proportionality constant is found by taking the weak field limit to be Newtonian gravity. We seek to derive these field equations.

There are two parts to general relativity. Most of general relativity is deduced from the equivalence principle (gravity bending light, gravitational time dilation,...); that's the first part. However, the second part, the field equations, are not deducible from the equivalence principle. Historically, Einstein guessed the field equations. His first two guesses were incorrect, but his third guess was the correct field equations presented above, (17.1) & (17.2). They are considered correct because they fit observation[62]. Einstein was guided to this correct guess by the fact that the Einstein tensor

---

[62] These are not the only field equations that fit observation. Cartan's torsoin theory of gravity and the Brans-Dicke scalar tensor theory of gravity also have field equations which match observation. Einstein's field equations are preferred because they are the simplest.

and the mass-energy tensor are the only two tensors around that have zero covariant divergence.

General relativity is based in a Riemann like geometric space. Such a space is constructed with:

a) A 4-dimensional manifold
b) An affine connection (notion of parallel transport of a vector)
c) A metric tensor which ensures that a vector has the same length everywhere in the manifold
d) Two types of 2-dimensional angles and the associated inner products

Can we reproduce this construction?

a) Adding the six A_3 algebras (taking the expectation algebra) produces a 4-dimensional manifold.
b) Allowing the phase (angle of a vector in the polar form) of an A_3 algebra to vary from point to point over the manifold induces an affine connection, a sense of parallel lines, into the manifold. Without a previously installed affine connection, it is meaningless to say that the direction of the A_3 vector varies from point to point in the manifold. The local variation of A_3 phase therefore induces an affine connection into the manifold.[63]
c) Insisting that the length of the A_3 vector, as defined by the expectation distance function $d^2 = t^2 - x^2 - y^2 - z^2$, is the same at all points in the manifold induces a metric tensor into the manifold.
d) The 2-dimensional expectation algebras are just the 2-dimensional algebras themselves because only one copy of each algebra derives from the finite group C_2. These 2-

---

[63] This procedure of allowing local phase variation over an underlying space is the source of the electromagnetic, weak, and strong forces in QFT. We are doing the same over a manifold rather than over an established space with an established affine connection. We assert that the procedure induces the affine connection into the manifold.

dimensional algebras hold invariant distance functions which are sub-functions of the 4-dimensional A_3 expectation distance function, and so we have 2-dimensional rotations in our 4-dimensional space-time. They bring with them the 2-dimensional inner products, curls etc..

We have reproduced the construction. We have produced our 4-dimensional space-time from the A_3 algebras.

The left-hand side of the GR field equations:
The first derivatives of the metric tensor are of no interest in Riemannian geometry. There is only one way to combine the six mixed second derivatives of the metric tensor together and produce a tensor. That tensor is the Riemann curvature tensor which encodes the curvature of the manifold. When we have the metric tensor of a manifold, we have the curvature tensor of the manifold.

Appropriately, contracting the curvature tensor and combining it with the metric tensor will give one of the two zero divergence tensors; this is the Einstein tensor, (17.1). We have the left-hand side of the field equations, (17.2), of general relativity.

The right-hand side of the GR field equations:
Gravity is similar to and different from electromagnetism. The Einstein tensor is a symmetric tensor with six independent components. The electromagnetic tensor is an anti-symmetric tensor with six independent components. There are two three component vector fields, the magnetic field and the electric field, within the electromagnetic tensor, but there is only one six component tensor gravitational field in the gravitation tensor.

The reader might think that, because we have had success in deriving the electromagnetic tensor by simply adding the quaternion and anti-quaternion $B \& E$ fields and because we have derived the distance function of the space-time in which we sit by simply adding the distance functions of the A_3 algebras, then we can use a similar technique to derive the mass-energy tensor from the A_3 algebras.

From simply adding the $B \& E$ fields of the six A_3 algebras we get a 'lop-sided' tensor:

$$SUM_{A_3}\left(E - Fields\right) =$$

$$4\begin{bmatrix} 3\left(\dfrac{\partial\phi}{\partial t}+\dfrac{\partial A_x}{\partial x}+\dfrac{\partial A_y}{\partial y}+\dfrac{\partial A_z}{\partial z}\right) & 3\dfrac{\partial A_x}{\partial t}+\dfrac{\partial\phi}{\partial x} & 3\dfrac{\partial A_y}{\partial t}+\dfrac{\partial\phi}{\partial y} & 3\dfrac{\partial A_z}{\partial t}+\dfrac{\partial\phi}{\partial z} \\[2mm] \dfrac{\partial A_x}{\partial t}+3\dfrac{\partial\phi}{\partial x} & \sim & 0 & 0 \\[2mm] \dfrac{\partial A_y}{\partial t}+3\dfrac{\partial\phi}{\partial y} & 0 & \sim & 0 \\[2mm] \dfrac{\partial A_z}{\partial t}+3\dfrac{\partial\phi}{\partial z} & 0 & 0 & \sim \end{bmatrix} \quad (17.3)$$

$$SUM_{A_3}\left(B - Fields\right) =$$

$$4\begin{bmatrix} 0 & 0 & 0 & 0 \\[2mm] 0 & 0 & 3\dfrac{\partial A_y}{\partial x}+\dfrac{\partial A_x}{\partial y} & 3\dfrac{\partial A_z}{\partial x}+\dfrac{\partial A_x}{\partial z} \\[2mm] 0 & \dfrac{\partial A_y}{\partial x}+3\dfrac{\partial A_x}{\partial y} & 0 & 3\dfrac{\partial A_z}{\partial y}+\dfrac{\partial A_y}{\partial z} \\[2mm] 0 & \dfrac{\partial A_z}{\partial x}+3\dfrac{\partial A_x}{\partial z} & \dfrac{\partial A_z}{\partial y}+3\dfrac{\partial A_y}{\partial z} & 0 \end{bmatrix} \quad (17.4)$$

This looks too messy to be correct. Perhaps we should try something else.

We take the above $B \& E$ fields and simply separate out the anti-symmetric parts of the tensors and the symmetric parts of the tensors giving the symmetric tensor:

$$SUM_{(A_3)} =$$

$$\frac{1}{4}\begin{bmatrix} 3\left(\frac{\partial\phi}{\partial t}+\frac{\partial A_x}{\partial x}+\frac{\partial A_y}{\partial y}+\frac{\partial A_z}{\partial z}\right) & 2\left(\frac{\partial A_x}{\partial t}+\frac{\partial\phi}{\partial x}\right) & 2\left(\frac{\partial A_y}{\partial t}+\frac{\partial\phi}{\partial y}\right) & 2\left(\frac{\partial A_z}{\partial t}+\frac{\partial\phi}{\partial z}\right) \\ 2\left(\frac{\partial A_x}{\partial t}+\frac{\partial\phi}{\partial x}\right) & \sim & 2\left(\frac{\partial A_y}{\partial x}+\frac{\partial A_x}{\partial y}\right) & 2\left(\frac{\partial A_z}{\partial x}+\frac{\partial A_x}{\partial z}\right) \\ 2\left(\frac{\partial A_y}{\partial t}+\frac{\partial\phi}{\partial y}\right) & 2\left(\frac{\partial A_y}{\partial x}+\frac{\partial A_x}{\partial y}\right) & \sim & 2\left(\frac{\partial A_z}{\partial y}+\frac{\partial A_y}{\partial z}\right) \\ 2\left(\frac{\partial A_z}{\partial t}+\frac{\partial\phi}{\partial z}\right) & 2\left(\frac{\partial A_z}{\partial x}+\frac{\partial A_x}{\partial z}\right) & 2\left(\frac{\partial A_z}{\partial y}+\frac{\partial A_y}{\partial z}\right) & \sim \end{bmatrix} \quad (17.5)$$

And the anti-symmetric tensor:

$$SUM_{[A_3]} =$$

$$\frac{1}{4}\begin{bmatrix} 0 & \left(\frac{\partial A_x}{\partial t}-\frac{\partial\phi}{\partial x}\right) & \left(\frac{\partial A_y}{\partial t}-\frac{\partial\phi}{\partial y}\right) & \left(\frac{\partial A_z}{\partial t}-\frac{\partial\phi}{\partial z}\right) \\ -\left(\frac{\partial A_x}{\partial t}-\frac{\partial\phi}{\partial x}\right) & 0 & \left(\frac{\partial A_y}{\partial x}-\frac{\partial A_x}{\partial y}\right) & \left(\frac{\partial A_z}{\partial x}-\frac{\partial A_x}{\partial z}\right) \\ -\left(\frac{\partial A_y}{\partial t}-\frac{\partial\phi}{\partial y}\right) & -\left(\frac{\partial A_y}{\partial x}-\frac{\partial A_x}{\partial y}\right) & 0 & \left(\frac{\partial A_z}{\partial y}-\frac{\partial A_y}{\partial z}\right) \\ -\left(\frac{\partial A_z}{\partial t}-\frac{\partial\phi}{\partial z}\right) & -\left(\frac{\partial A_z}{\partial x}-\frac{\partial A_x}{\partial z}\right) & -\left(\frac{\partial A_z}{\partial y}-\frac{\partial A_y}{\partial z}\right) & 0 \end{bmatrix} \quad (17.6)$$

Note that by choosing conjugate potentials, the anti-symmetric tensor can be made to match the electromagnetic tensor.

We assert that the symmetric tensor might be the mass-energy tensor on the right-hand side of (17.2). If it is, then we would have force dependent upon mass-energy. Perhaps it is that simple and we should not think of electromagnetism as a purely quaternion phenomenon but think of it instead as an anti-symmetric phenomenon[64]; to match

---

[64] Perhaps there are two parts to the electric charge coming from the quaternions and the A_3 algebras, and perhaps these are weak isospin and weak hypercharge connected by something like the Gell-Mann-Nishijima relation.

this, we might think of gravity as a symmetric phenomenon. Thus, both gravity and electromagnetism are within the $C_2 \times C_2$ group.

Taking the symmetric tensor (17.5) to be the mass-energy tensor and 'guessing together' as Einstein did this with the derived left-hand side of (17.2), we have the field equations of general relativity.

Although we do not yet see it, perhaps there is something we have missed that avoids our having to emulate Einstein and guess the field equations. Perhaps they can be deduced. None-the-less, we have the field equations of general relativity as well as Einstein had them, but we also have the 4-dimensional space-time of our universe.

We have unification between gravity and electromagnetism.

Another source of curvature perhaps:

Above, we have used the Cartesian forms of the A_3 algebras, but we know that only restricted forms of these Cartesian presentations are *bona fide* division algebras. The A_3 algebras ought to be presented in their polar form as a real radial variable and a rotation matrix. A rotation matrix maps out a spherical[65] $(n-1)$-dimensional 'surface' of unit distance from the origin; the radial variable just moves this spherical 'surface' away from or closer to the origin and so expands or shrinks the $(n-1)$-dimensional 'area' of this spherical 'surface'. Such a $(n-1)$-dimensional surface is a group of infinite order whose group operation is multiplication by the rotation matrix. Within the $\{\mathbb{R}^n, \mathbb{C}^n, \mathbb{H}^n, \mathbb{O}^n\}$ spaces, we call such groups Lie groups[66]. We think of a Lie group as a curved sub-space embedded in a higher dimensional space. Within an A_3 algebra, for a stationary observer, the real variable is the time variable. Since we cannot move

---

[65] This includes hyperbolic spherical.

[66] The term 'Lie group' normally refers to spherical surfaces in $\mathbb{R}^n, \mathbb{C}^n, \mathbb{H}^n, \mathbb{O}^n$, but the idea also applies to division algebra spaces. In division algebra spaces, a Lie group is no more than a rotation matrix.

backwards and forwards along the time axis at will, in an A_3 algebra, we find ourselves stuck in a spherical surface of distance t from the origin. Such a spherical surface has a constant curvature. Note that this is a 3-dimensional curved surface embedded within a 4-dimensional space-time. We thus opine that we, as stationary observers, are sitting in six A_3 Lie groups each of which has a constant curvature. If we allow that the curvatures of these six Lie groups need not be equal and that they can vary from point to point in space-time, we are sitting in a 4-dimensional space-time that is described by six curvatures. There are six principle curvatures in a 4-dimensional Riemannian type space. Perhaps this is the source of our affine connection rather than a locally varying A_3 phase.

Gravito-e/magnetism (GEM) in the A_3 algebras:

If we can get gravity from aggregating the A_3 algebras as we have above, what are we to do with the Maxwell equations of the individual A_3 algebras; are they not some kind of force over space-time? There might be an answer, and we now look at this. Of course, we opine that the A_3 Maxwell equations within only one A_3 algebra are a quantum phenomenon.

There is a somewhat esoteric area of physics called gravito-electromagnetism (GEM)[67]. GEM is concerned, along with other things, with the Lense-Thirring effect presently being tested by the satellite Gravity Probe B (also called the Stanford Gyroscope Experiment). GEM is thought to be connected to quantum gravity. GEM is often, in the opinion of some wrongly[68], referred to as 'Frame Dragging'. GEM is developed as an analogy to electromagnetism that *"...supplies us with a familiar model for the extra gravitational force*

---

[67] A summary of GEM is offered in the paper Gravitoelectromagnetism: A Brief Review, Bahram Mashhoon arXiv:gr-qc/0311030v2 17th April 2008.
[68] W. Rindler, The case against space dragging. Phys. Lett. A233, 25 (1997)

created by the motion of the sources and felt only by moving particles..."[69].

Traditionally, GEM is seen as an approximation to general relativity and is based upon linearised GR equations, however, it can be treated exactly. The exact treatment takes a vector potential, A_μ, and defines two sectors of its first derivatives as:

$$F_{\mu\nu} = \partial_\mu A_\nu - \partial_\nu A_\mu$$
$$H_{\mu\nu} = \partial_\mu A_\nu + \partial_\nu A_\mu$$
(17.7)

In this book, we would call these the conventional curl and the space-time curl. In GEM, there are two fields analogous to the electric and magnetic fields of electromagnetism. There are also GEM Maxwell equations. We list them alongside the electromagnetic Maxwell equations:

$$\nabla \cdot E_g = -4\pi G \rho_g \qquad\qquad \nabla \cdot E = \frac{1}{\varepsilon_0}\rho$$

$$\nabla \cdot B_g = 0 \qquad\qquad \nabla \cdot B = 0$$

(17.8)

$$\nabla \times E_g = -\frac{\partial B_g}{\partial t} \qquad\qquad \nabla \times E = -\frac{\partial B}{\partial t}$$

$$\nabla \times B_g = 4\left(-\frac{4\pi G}{c^2}J_g + \frac{1}{c^2}\frac{\partial E_g}{\partial t}\right) \qquad\qquad \nabla \times B = \frac{1}{\varepsilon_0 c^2}J + \frac{1}{c^2}\frac{\partial E}{\partial t}$$

Within this conventional listing, (17.8), the curls and divergences are of the conventional forms that derive from the wholly anti-symmetric algebras such as the quaternions. We have seen above, (5.18), that space-time curls are different from the conventional curl. Of course, the A_3 algebras each have their own type of curl and their own set of Maxwell equations.

---

[69] W. Rindler, Relativity, Special, General and Cosmological. Pg 336. Oxford Uni. Press 0-19-850836-0

Looking at the electromagnetic tensor above, we might take a single A_3 space to be a quantum GEM space. We take a single A_3 algebra to be a potential.

$$\Phi_{SSA} = \begin{bmatrix} \phi & A_x & A_y & A_z \\ A_x & \phi & A_z & A_y \\ A_y & -A_z & \phi & -A_x \\ -A_z & A_y & -A_x & \phi \end{bmatrix}$$

$$\{\phi, A_i\} = f(t, x, y, z)$$

(17.9)

Non-commutative differentiation leads to the $E \& B$ fields:

$$E_{[1,1]} = \frac{\partial \phi}{\partial t} + \frac{\partial A_x}{\partial x} + \frac{\partial A_y}{\partial y} + \frac{\partial A_z}{\partial z}$$

$$E_{[1,2]} = \frac{\partial A_x}{\partial t} + \frac{\partial \phi}{\partial x}$$

$$E_{[1,3]} = \frac{\partial A_y}{\partial t} + \frac{\partial \phi}{\partial y}$$

$$E_{[1,4]} = \frac{\partial A_z}{\partial t} - \frac{\partial \phi}{\partial z}$$

(17.10)

and:

$$B_{[1,1]} = 0$$

$$B_{[1,2]} = -\left(\frac{\partial A_z}{\partial y} + \frac{\partial A_y}{\partial z} \right)$$

(17.11)

$$B_{[1,3]} = \frac{\partial A_x}{\partial z} + \frac{\partial A_z}{\partial x}$$

$$B_{[1,4]} = \frac{\partial A_y}{\partial x} - \frac{\partial A_x}{\partial y}$$

(17.12)

These are of the form:

$$\begin{bmatrix} E_t & E_x & E_y & E_z \\ E_x & E_t & E_z & E_y \\ E_y & -E_z & E_t & -E_x \\ -E_z & E_y & -E_x & E_t \end{bmatrix} \quad \& \quad \begin{bmatrix} B_t & B_x & B_y & B_z \\ B_x & B_t & B_z & B_y \\ B_y & -B_z & B_t & -B_x \\ -B_z & B_y & -B_x & B_t \end{bmatrix} \quad (17.13)$$

We think these might be our quantum GEM fields.

Gravito-electromagnetic Maxwell equations:
The differential identity that is the homogeneous *SSA* Maxwell equations is:

$$\{B,d\}_{SSA} = [E,d]_{SSA}$$

$$\frac{\partial B_x}{\partial x} + \frac{\partial B_y}{\partial y} + \frac{\partial B_z}{\partial z} = 0$$

$$\frac{\partial B_x}{\partial t} = -\left(\frac{\partial E_z}{\partial y} + \frac{\partial E_y}{\partial z} \right) \qquad (17.14)$$

$$\frac{\partial B_y}{\partial t} = \frac{\partial E_x}{\partial z} + \frac{\partial E_z}{\partial x}$$

$$\frac{\partial B_z}{\partial t} = \frac{\partial E_y}{\partial x} - \frac{\partial E_x}{\partial y}$$

We see that, subject to the type of curl, we get something like the GEM fields and the GEM Maxwell equations. Clearly, we can swap a few signs about by taking a conjugate potential, but we could not match the conventional curl used in the conventional GEM Maxwell equations. If we took the SSA_{Anti} algebra, we would get some kind of anti-matter equivalent with reversed B field.

$$\{B,d\}_{SSA_{anti}} = [E,d]_{SSA_{anti}} \qquad (17.15)$$

$$\frac{\partial B_x}{\partial x} + \frac{\partial B_y}{\partial y} + \frac{\partial B_z}{\partial z} = 0 \qquad (17.16)$$

$$\frac{\partial B_x}{\partial t} = \frac{\partial E_z}{\partial y} + \frac{\partial E_y}{\partial z}$$

$$\frac{\partial B_y}{\partial t} = -\left(\frac{\partial E_x}{\partial z} + \frac{\partial E_z}{\partial x}\right) \qquad (17.17)$$

$$\frac{\partial B_z}{\partial t} = -\frac{\partial E_y}{\partial x} + \frac{\partial E_x}{\partial y}$$

The standard form of the GEM Maxwell equations is not based on observation because the GEM fields are too weak to be easily observed separate from background gravitational effects. As mentioned above, these observations might be available within a few years. The GEM Maxwell equations are simply assumed to match the electromagnetic Maxwell equations. It might be that, when gravitational variables are substituted for electromagnetic variables, the A_3 type of curl should also be substituted for the quaternion type of curl. Perhaps the observations will reveal that the A_3 Maxwell equations are of the correct quantitised GEM form and that the A_3 tensor formed by summing an A_3 algebra and its anti-algebra is the correct macroscopic GEM form.

Three generations of particles?:

We seem to have discovered gravity within the A_3 algebras. Gravity is associated with mass, and so we think that the A_3 algebras have a charge that is mass. There are three pairs of A_3 algebras. If each pair of A_3 algebras is associated with mass in a way analogous to the way quaternion algebras are associated with electric charge, we will have three separate 'mass charges'. Presumably each particle in the universe would interact with each A_3 algebra in a similar way, and so each particle would have three masses. That is, we would have three generations of particles.

Chapter 18

Non-Commutative Differentiation Operators

We have the $SU(2)$ differentiation operator:

$$\partial Q_{(t,x,y,z)} = \begin{bmatrix} \dfrac{\partial}{\partial t} & -\dfrac{\partial}{\partial x} & -\dfrac{\partial}{\partial y} & -\dfrac{\partial}{\partial z} \\[2mm] \dfrac{\partial}{\partial x} & \dfrac{\partial}{\partial t} & \dfrac{\partial}{\partial z} & -\dfrac{\partial}{\partial y} \\[2mm] \dfrac{\partial}{\partial y} & -\dfrac{\partial}{\partial z} & \dfrac{\partial}{\partial t} & \dfrac{\partial}{\partial x} \\[2mm] \dfrac{\partial}{\partial z} & \dfrac{\partial}{\partial y} & -\dfrac{\partial}{\partial x} & \dfrac{\partial}{\partial t} \end{bmatrix} \qquad (18.1)$$

The matrix product of this differentiation operator and a quaternion potential will give the left and right differentials of the quaternion potential. We can differentiate non-commutatively by simply multiplying the differentiation operator by the potential to the left or to the right. The $B \& E$ fields are:

$$E = \frac{1}{2}\left(d_L + d_R\right)$$

$$= \frac{1}{2}\left(\partial Q\Phi + \Phi\partial Q\right)$$

$$B = \frac{1}{2}\left(d_L - d_R\right) \qquad (18.2)$$

$$= \frac{1}{2}\left(\partial Q\Phi - \Phi\partial Q\right)$$

Conjugate potential:

We differentiate the quaternion conjugate potential by simply putting it in the place of the quaternion potential but still using the same $SU(2)$ differential operator.

A_3 Differential operators:

The A_3 algebras contain symmetric imaginary variables, and the symmetric differentials are their own inverses, and the differential operator is the sum of these inverses. For example, the differential operator of the SSA1 algebra:

$$\begin{bmatrix} \phi & A_x & A_y & A_z \\ A_x & \phi & A_z & A_y \\ A_y & -A_z & \phi & -A_x \\ -A_z & A_y & -A_x & \phi \end{bmatrix} \tag{18.3}$$

is:

$$\begin{bmatrix} \partial t & \partial x & \partial y & -\partial z \\ \partial x & \partial t & -\partial z & \partial y \\ \partial y & \partial z & \partial t & -\partial x \\ \partial z & \partial y & -\partial x & \partial t \end{bmatrix} \tag{18.4}$$

Which is not the conjugate algebra.

Double differentiation:

Double differentiation is:

$$d_L d_L = \partial Q \partial Q (\Phi)$$
$$d_R d_R = \Phi \partial Q \partial Q$$
$$d_R d_L = \partial Q \Phi \partial Q \tag{18.5}$$
$$d_L d_R = \partial Q \Phi \partial Q$$

Within which, the fundamental identity $d_L d_R = d_R d_L$ is obvious. In terms of the $B \& E$ fields, the other fundamental identity is:

$$\{B,d\} = [E,d]$$
$$\partial QB + B\partial Q = \partial QE - E\partial Q$$

(18.6)

Momentum operators:

Each imaginary variable within a quaternion, when taken with the real variable, forms a, 4-dimensional 2-dimensional (sorry again), $\mathbb{C}$ sub-algebra. We saw earlier that the quantum mechanical momentum operator is just differentiate with respect to the imaginary variable within the $\mathbb{C}$ algebra and that, within that algebra, $\hbar = \dfrac{1}{\lambda}$; where λ is a scaling parameter.

A scaled quaternion is a $C_2 \times C_2$ non-commutative algebra in which the scaling factors are such that $\{\alpha < 0, \eta < 0, \varepsilon > 0\}$. We accommodate these by setting all these scaling parameters to be positive but placing minus signs in the appropriate places within the matrix as is required to have the quaternion algebra. This produces a recognisable quaternion matrix (distribution of the minus signs) but still a scaled algebra. We will differentiate a scaled quaternion with respect to the d variable. We have the left differential:

$$
\partial \begin{bmatrix} f & 0 & - & j \\ - & f & -\dfrac{\alpha}{\varepsilon}j & - \\ - & \dfrac{\eta}{\varepsilon}j & f & - \\ -\dfrac{\alpha\eta}{\varepsilon^2}j & 0 & 0 & f \end{bmatrix}
$$

$$
\partial \begin{bmatrix} 0 & 0 & 0 & d \\ 0 & 0 & -\dfrac{\alpha}{\varepsilon}d & 0 \\ 0 & \dfrac{\eta}{\varepsilon}d & 0 & 0 \\ -\dfrac{\alpha\eta}{\varepsilon^2}d & 0 & 0 & 0 \end{bmatrix}
$$

$$
= \frac{1}{\begin{bmatrix} 0 & 0 & 0 & 1 \\ 0 & 0 & -\dfrac{\alpha}{\varepsilon} & 0 \\ 0 & \dfrac{\eta}{\varepsilon} & 0 & 0 \\ -\dfrac{\alpha\eta}{\varepsilon^2} & 0 & 0 & 0 \end{bmatrix}} \begin{bmatrix} \dfrac{\partial f}{\partial d} & 0 & 0 & \dfrac{\partial j}{\partial d} \\ 0 & \dfrac{\partial f}{\partial d} & -\dfrac{\alpha}{\varepsilon}\dfrac{\partial j}{\partial d} & 0 \\ 0 & \dfrac{\eta}{\varepsilon}\dfrac{\partial j}{\partial d} & \dfrac{\partial f}{\partial d} & 0 \\ -\dfrac{\alpha\eta}{\varepsilon^2}\dfrac{\partial j}{\partial d} & 0 & 0 & \dfrac{\partial f}{\partial d} \end{bmatrix} \quad (18.7)
$$

The left-most factor is:

$$
\begin{bmatrix}
0 & 0 & 0 & 1 \\
0 & 0 & -\dfrac{\alpha}{\varepsilon} & 0 \\
0 & \dfrac{\eta}{\varepsilon} & 0 & 0 \\
-\dfrac{\alpha\eta}{\varepsilon^2} & 0 & 0 & 0
\end{bmatrix}^{-1}
$$

$$
=
\begin{bmatrix}
-\dfrac{\varepsilon^2}{\alpha\eta} & 0 & 0 & 0 \\
0 & -\dfrac{\varepsilon^2}{\alpha\eta} & 0 & 0 \\
0 & 0 & -\dfrac{\varepsilon^2}{\alpha\eta} & 0 \\
0 & 0 & 0 & -\dfrac{\varepsilon^2}{\alpha\eta}
\end{bmatrix}
\begin{bmatrix}
0 & 0 & 0 & 1 \\
0 & 0 & -\dfrac{\alpha}{\varepsilon} & 0 \\
0 & \dfrac{\eta}{\varepsilon} & 0 & 0 \\
-\dfrac{\alpha\eta}{\varepsilon^2} & 0 & 0 & 0
\end{bmatrix}
$$

$$
\equiv -k_\lambda
\begin{bmatrix}
\dfrac{\varepsilon^2}{\alpha\eta} & 0 & 0 & 0 \\
0 & \dfrac{\varepsilon^2}{\alpha\eta} & 0 & 0 \\
0 & 0 & \dfrac{\varepsilon^2}{\alpha\eta} & 0 \\
0 & 0 & 0 & \dfrac{\varepsilon^2}{\alpha\eta}
\end{bmatrix}
\tag{18.8}
$$

We have subscripted a λ to the $\hat{k}$ to indicate this is the scaled version. We put a little hat on the quaternion imaginary units so they look prettier than their counterpart in the $\mathbb{C}$ algebra. We have

$$
\hbar_z = \frac{\varepsilon^2}{\alpha\eta}
\tag{18.9}
$$

We have subscripted a z to the aitch-bar. We again, obviously, have the linear momentum operator:

$$p_z = -k\hbar_z \frac{\partial}{\partial z} \qquad (18.10)$$

Of course, if we had taken the right differential, we would have come to the same answer because this 2-dimensional sub-algebra, $\mathbb{C}$, is a commutative algebra. Similarly we have:

$$\hbar_x = \frac{1}{\alpha} \qquad p_x = -\hat{i}\hbar_x \frac{\partial}{\partial x} \qquad (18.11)$$

And:

$$\hbar_y = \frac{1}{\eta} \qquad p_y = -j\hbar_y \frac{\partial}{\partial y} \qquad (18.12)$$

Of course, in an equally scaled space, $\hbar_x = \hbar_y = \hbar_z$. Well, the above is very nice, but we ought to deal with the quaternions as a 4-dimensional algebra rather than as three 2-dimensional algebras. Indeed, we are probably[70] out of order to separate the quaternion into 2-dimensional sub-algebras even if we do acknowledge that they are 4-dimensional 2-dimensional sub-algebras.

Angular momentum operators:
Within quantum mechanics, we have, based upon the correspondence principle, the angular momentum operators as:

$$L_x = y p_z - z p_y \qquad L_y = z p_x - x p_z$$
$$L_z = x p_y - y p_x \qquad (18.13)$$

These are:

---

[70] Your author is unsure about this.

$$L_x = -i\hbar \left(y\frac{\partial}{\partial z} - z\frac{\partial}{\partial y} \right)$$

$$L_y = -i\hbar \left(z\frac{\partial}{\partial x} - x\frac{\partial}{\partial z} \right) \qquad (18.14)$$

$$L_z = -i\hbar \left(x\frac{\partial}{\partial y} - y\frac{\partial}{\partial x} \right)$$

The quaternion differential operator (the $SU(2)$ differential operator) acting upon a quaternion is:

$$\partial Q(Q) = PROD \left(\begin{bmatrix} \dfrac{\partial}{\partial t} & -\dfrac{\partial}{\partial x} & -\dfrac{\partial}{\partial y} & -\dfrac{\partial}{\partial z} \\[2mm] -\alpha\dfrac{\partial}{\partial x} & \dfrac{\partial}{\partial t} & -\dfrac{\alpha}{\varepsilon}\dfrac{\partial}{\partial z} & \varepsilon\dfrac{\partial}{\partial y} \\[2mm] -\eta\dfrac{\partial}{\partial y} & \dfrac{\eta}{\varepsilon}\dfrac{\partial}{\partial z} & \dfrac{\partial}{\partial t} & -\varepsilon\dfrac{\partial}{\partial x} \\[2mm] -\dfrac{\alpha\eta}{\varepsilon^2}\dfrac{\partial}{\partial z} & -\dfrac{\eta}{\varepsilon}\dfrac{\partial}{\partial y} & \dfrac{\alpha}{\varepsilon}\dfrac{\partial}{\partial x} & \dfrac{\partial}{\partial t} \end{bmatrix} \begin{bmatrix} t & x & y & z \\[2mm] -\alpha x & t & -\dfrac{\alpha}{\varepsilon}z & \varepsilon y \\[2mm] -\eta y & \dfrac{\eta}{\varepsilon}z & t & -\varepsilon x \\[2mm] -\dfrac{\alpha\eta}{\varepsilon^2}z & -\dfrac{\eta}{\varepsilon}y & \dfrac{\alpha}{\varepsilon}x & t \end{bmatrix} \right) \qquad (18.15)$$

Of course, we have the right differential $(Q)\partial Q$ also. This leads to the $B \& E$ fields. The B field is:

$$B_{[1,1]} = 0$$

$$B_{[1,2]} = -\frac{\eta}{\varepsilon}\left(y\frac{\partial}{\partial z} - z\frac{\partial}{\partial y} \right)$$

$$B_{[1,3]} = -\frac{\alpha}{\varepsilon}\left(z\frac{\partial}{\partial x} - x\frac{\partial}{\partial z} \right) \qquad (18.16)$$

$$B_{[1,4]} = -\varepsilon\left(x\frac{\partial}{\partial y} - y\frac{\partial}{\partial x} \right)$$

Of course, the positions in the quaternion matrix correspond to the $\{\hat{i}, j, k\}$, and so we can write the B field in non-matrix notation as:

$$B_x = -\frac{\eta}{\varepsilon}\hat{i}\left(y\frac{\partial}{\partial z} - z\frac{\partial}{\partial y} \right)$$

$$B_y = -\frac{\alpha}{\varepsilon}j\left(z\frac{\partial}{\partial x} - x\frac{\partial}{\partial z} \right) \quad : \quad \{\hat{i}, j, k\} = \sqrt{-1} \qquad (18.17)$$

$$B_z = -\varepsilon k\left(x\frac{\partial}{\partial y} - y\frac{\partial}{\partial x} \right)$$

Wherein we see that the scaling parameters have now taken the form:

$$\hbar_x = \frac{\eta}{\varepsilon}, \quad \hbar_y = \frac{\alpha}{\varepsilon}, \quad \hbar_z = \varepsilon \qquad (18.18)$$

Of course, if $\alpha = \varepsilon = \eta = 1$, the $\hbar$ s are all the same; and so, it seems that we have the quantum mechanical angular momentum operators as the B field of the $SU(2)$ differential operator. There might be some deep connection here to the rotational nature of the magnetic field that we have identified as the quaternion B field.

What about the E field? Using the $SU(2)$ differential operator and a quaternion as above, the E field is:

$$E_{[1,1]} = -\alpha x \frac{\partial}{\partial x} - \eta y \frac{\partial}{\partial y} - \frac{\alpha \eta}{\varepsilon^2} z \frac{\partial}{\partial z}$$

$$E_{[1,2]} = x \frac{\partial}{\partial t} + t \frac{\partial}{\partial x}$$

$$E_{[1,3]} = y \frac{\partial}{\partial t} + t \frac{\partial}{\partial y} \qquad (18.19)$$

$$E_{[1,4]} = z \frac{\partial}{\partial t} + t \frac{\partial}{\partial z}$$

Your author has no idea what to make of this E field. Actually, in spite of having spent his time producing this chapter, your author does not like the whole concept of differential operators or any other operators and would be happier if operators were kicked out of quantum mechanics all together.

Chapter 19

Measuring Non-commutativity

It is an established fact that for matrices $\{A, B\}$, if the matrices commute, we have:

$$e^A e^B = e^{A+B} \qquad (19.1)$$

and that, if the matrices do not commute, in general:

$$e^A e^B \neq e^{A+B} \qquad (19.2)$$

Within division algebras, we get rotation matrices by taking the exponential of the matrix of the imaginary variables (zero on the leading diagonal). Further, the $(n-1)$ imaginary variables become the $(n-1)$ variables that are the n-dimensional angle within the rotation matrix. Clearly (multiplicative closure of form), the rotation matrices of commutative algebras are commutative and the rotation matrices of non-commutative algebras are non-commutative. It follows from (19.1) that, in a commutative algebra, a rotation through an angle, say, (b, c) followed by a rotation through an angle, say, (f, g) is a rotation through the angle $(b+f, c+g)$. This is exactly what (19.1) means. This is exactly to what we are accustomed and what we have with the Euclidean circle, of course. It similarly follows from (19.2) that, in a non-commutative algebra, a rotation through an angle, say, (b, c, d) followed by a rotation through an angle, say, (f, g, h) is generally not a rotation through the angle $(b+f, c+g, d+h)$. This is exactly what (19.2) means. We might even define commutative and non-commutative rotation by this product of rotations is the sum of angles or not the sum of the angles property.

(19.2) does not say that the angle differs with the order of multiplication. In fact the angle in a product of two non-commutative rotations is the same for both orders of multiplication of the rotation matrices. The non-commutativity of the rotations is not in the angle. When we multiply two non-commutative rotation matrices together, the leading diagonal element of the product, the leading trigonometric function, is the same regardless of the order of multiplication. If the non-commutativity was in the angle, this could not be the case. Nor is the non-commutativity within the trigonometric functions. Of course, the angle within the leading trigonometric function is the same angle as we have in the other trigonometric functions within a rotation matrix. It is just that, in the non-commutative case, the angle is not simply a sum of the angular variables in the factor matrices. The non-commutativity is in the form of the matrix and not in the angle – we need matrices to have non-commutativity.

In three dimensions, because the C_3 algebras are commutative, we have:

$$e^{\begin{bmatrix} 0 & b & c \\ c & 0 & b \\ b & c & 0 \end{bmatrix}} e^{\begin{bmatrix} 0 & f & g \\ g & 0 & f \\ f & g & 0 \end{bmatrix}} = e^{\begin{bmatrix} 0 & b+f & c+g \\ c+g & 0 & b+f \\ b+f & c+g & 0 \end{bmatrix}} \tag{19.3}$$

and so, within the C_3 algebras, the angle in the product of two rotation matrices is a simple sum of the angles in the factor matrices. Multiplication of two commutative rotation matrices, in either order, will then lead to addition relations between the trigonometric functions of the particular algebra. For the above algebra, in which the three trigonometric functions are $\{v_A, v_B, v_C\}$, we have:

$$v_A(b+e, c+f) = v_A(b,c)v_A(e,f) + v_B(b,c)v_C(e,f) \\ + v_C(b,c)v_B(e,f) \tag{19.4}$$

$$v_B(b+e, c+f) = v_A(b,c)v_B(e,f) + v_B(b,c)v_A(e,f) \\ + v_C(b,c)v_C(e,f) \tag{19.5}$$

$$v_C(b+e,c+f) = v_A(b,c)v_C(e,f) + v_B(b,c)v_B(e,f)$$
$$+ v_C(b,c)v_A(e,f) \tag{19.6}$$

These are just the 3-dimensional analogue of 2-dimensional addition relations like:

$$\cos(a+b) = \cos(a)\cos(b) - \sin(a)\sin(b)$$
$$\sin(a+b) = \cos(a)\sin(b) + \sin(a)\cos(b) \tag{19.7}$$

For non-commutative algebras, we will not have such trigonometric addition relations.

Measuring the amount of non-commutativity:
Even non-commutative rotation matrices commute sometimes. If both non-commutative rotations are in the same 2-dimensional plane[71], then, because all 2-dimensional rotations are commutative[72], these particular non-commutative rotations will be commutative. When two non-commutative rotations are commutative, the leading trigonometric function of the product of those two non-commutative rotations will be equal to the leading trigonometric function of the algebra with an angle that is the sum of the angular variables as in (19.1). We can measure the amount of non-commutativity as the difference between the leading trigonometric function in the non-commutative product and this leading trigonometric function with the sum of variables as its angle. This needs an example for clarification.

We will use quaternion angles as examples of 4-dimensional non-commutative angles. The quaternion trigonometric functions are:

$$v_A = \cos\left(\sqrt{b^2 + c^2 + d^2}\right) \tag{19.8}$$

---

[71] Higher dimensional 2-dimensional planes do not necessarily exist in higher dimensional algebras, but they do exist in the $C_2 \times C_2 \times \ldots$ algebras.

[72] Even 4-dimensional 2-dimensional rotations are commutative.

$$V_B = \frac{b}{\sqrt{b^2 + c^2 + d^2}} \sin\left(\sqrt{b^2 + c^2 + d^2}\right)$$

$$V_C = \frac{c}{\sqrt{b^2 + c^2 + d^2}} \sin\left(\sqrt{b^2 + c^2 + d^2}\right) \quad (19.9)$$

$$V_D = \frac{d}{\sqrt{b^2 + c^2 + d^2}} \sin\left(\sqrt{b^2 + c^2 + d^2}\right)$$

We have the quaternion:

$$\mathbb{H} = \begin{bmatrix} a & b+f & c+g & d+h \\ -b-f & a & -d-h & c+g \\ -c-g & d+h & a & -b-f \\ -d-h & -c-g & b+f & a \end{bmatrix} \quad (19.10)$$

Taking the exponential of the imaginary part of (19.10), $a = 0$, leads to the quaternion rotation matrix:

$$\begin{bmatrix} \cos\left(\sqrt{(b+f)^2 + (c+g)^2 + (d+h)^2}\right) & \sim & \sim & \sim \\ & \sim & & \sim & \sim & \sim \\ & \sim & & & \sim & \sim & \sim \\ & \sim & & & \sim & \sim & \sim \end{bmatrix} \quad (19.11)$$

This is (part of) rotation through the 4-dimensional angle $(b+f, c+g, d+h)$. The leading trigonometric function of the quaternion algebra with this angle is:

$$\cos\left(\sqrt{(b+f)^2 + (c+g)^2 + (d+h)^2}\right) \quad (19.12)$$

It is against this that we are going to compare the leading diagonal element of the product of two quaternion matrices to measure the amount of non-commutativity.

We have that the product of two quaternion rotations gives the leading diagonal element of the product matrix as:

$$\left[ROT\left(b,c,d\right) ROT\left(f,g,h\right)\right]_{[1,1]} =$$

$$\cos\left(\sqrt{b^2 +c^2 +d^2}\right)\cos\left(\sqrt{f^2 +g^2 +h^2}\right) \qquad (19.13)$$

$$-\frac{\left(bf +cg + dh\right)\sin\left(\sqrt{b^2 +c^2 +d^2}\right)\sin\left(\sqrt{f^2 +g^2 +h^2}\right)}{\sqrt{b^2 +c^2 +d^2}\sqrt{f^2 +g^2 +h^2}}$$

This does not change with the order of the factor matrices because it is an element on the leading diagonal of the product matrix.

We measure the amount of non-commutativity as the difference between (19.13) and (19.12).

$$\cos\left(\sqrt{\left(b+f\right)^2 +\left(c+g\right)^2 +\left(d+h\right)^2}\right)-$$

$$\cos\left(\sqrt{b^2 +c^2 +d^2}\right)\cos\left(\sqrt{f^2 +g^2 +h^2}\right)+ \qquad (19.14)$$

$$\frac{\left(bf +cg + dh\right)\sin\left(\sqrt{b^2 +c^2 +d^2}\right)\sin\left(\sqrt{f^2 +g^2 +h^2}\right)}{\sqrt{b^2 +c^2 +d^2}\sqrt{f^2 +g^2 +h^2}} = \Omega$$

Wherein we have used Ω to symbolise the amount of non-commutativity[73]. When this difference is zero, the rotations are commutative.

When all but one of the arguments in the 4-dimensional angles are zero, the above, (19.13), reduces to:

$$\cos\left(b+f\right) = \cos\left(b\right)\cos\left(f\right) - \sin\left(b\right)\sin\left(f\right) \qquad (19.15)$$

---

[73] When one introduces a symbol for some mathematical entity that has never before been symbolised, one is conscious of the possibility that mathematicians might be stuck with this symbol for centuries; the use of i to indicate the square root of minus one is an example. One does not want to use a very esoteric symbol that other mathematicians might not find in their computer, but nor does one want to use a symbol that is used elsewhere and whose usage might lead to confusion. One does one's best and one apologises if one has chosen unwisely.

We know this is correct. This is just another way of saying that 2-dimensional rotations are commutative.

With $c = d = f = g = h = 0.5$, we get:

$$\Omega = \cos\left(\sqrt{(b+0.5)^2 + 2}\right) - \cos\left(\sqrt{b^2 + 0.5}\right)\cos\left(\sqrt{0.75}\right)$$

$$+ \frac{(0.5b + 0.5)\sin\left(\sqrt{b^2 + 0.5}\right)\sin\left(\sqrt{0.75}\right)}{\sqrt{b^2 + 0.5}\sqrt{0.75}} \quad (19.16)$$

We have put definite values on so many of the variables because we cannot display a 4-dimensional graph. If the rotations were commutative, (19.16) would be zero for all values of b. The actual graph for different values of b, regardless of order of multiplication of course, is:

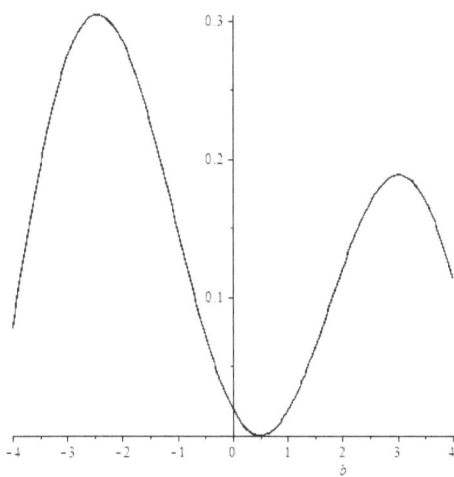

or

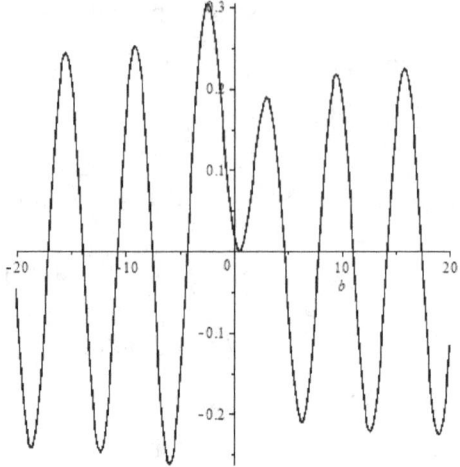

Other than at near $b = 0$, this seems to be a sine wave. At least in this case, the amount of non-commutativity is never infinite. The zeros of the graph are where the rotations are commutative. One wonders if one should take the square of Ω as the measure of non-commutativity.

Chapter 20

Lie Algebra from a Different Perspective

Conventional (classical) Lie algebras live in linear spaces. Such linear spaces are inventions of mathematicians who fit together n 1-dimensional spaces at right angles to each other and impose an invented distance function on those linear spaces. Sometimes, such an invented linear space coincides with the spatial structure of a division algebra because each division algebra is also a linear space.

We expect that, since we are working in division algebra spaces rather than the $\{\mathbb{R}^n, \mathbb{C}^n, \mathbb{H}^n, \mathbb{O}^n\}$ types of space, we will have different types of Lie groups; that is different types of 'spherical surfaces' at unit distance from the origin. We do find, surprisingly, both $SU(2)$ and parts of $SO(3,1)$ within our division algebra types of space, but, other than these two standard Lie groups, it seems there are no more coincidences. We opine that standard Lie groups are 'spherical surfaces' within the $\{\mathbb{R}^n, \mathbb{C}^n, \mathbb{H}^n, \mathbb{O}^n\}$ spaces. Except in the 2-dimensional cases, these do not correspond to rotation matrices. There are 'spherical surfaces' in division algebra spaces. Thus there are 'Lie groups' within the division algebras. The division algebra 'spherical surfaces' or 'Lie groups' all correspond to rotation matrices.

The particle content of the universe:
Within the standard model of particle physics, it is assumed that each generator of a (standard) Lie group corresponds to a boson. Thus, we have three bosons $\{W^\pm, Z^0\}$ associated with $SU(2)$ and eight gluons associated with the eight generators of $SU(3)$. The standard model does not predict the fermions. If we apply a similar correspondence

between non-commutative generators and bosons to the division algebra Lie groups and we add to this a correspondence between commutative generators and fermions, other than the Higgs boson, we get a match with the particle content of the universe provided we accept that there are only six gluons because there are only six things for gluons to do. This assumes the strong force is associated with the 8-dimensional algebras that have six non-commutative imaginary variables and one commutative imaginary variable and the commutative real variable.

Lie algebra and thank you Howard Georgi:
Howard Georgi produced a standard text book on Lie algebras. We will now quote extensively from Howard Georgi[74]. We will do this to compare the standard Lie algebra to the Lie algebra of division algebras. We begin.

"Suppose our group elements $g \in G$ depend smoothly on a continuous set of parameters

$$g(\alpha) \qquad (20.1)"$$

The group G to which Georgi refers is a finite group like C_3 or S_3 or even $C_2 \times C_2$. Both standard Lie algebra and our Lie algebra derive from the finite groups. At first, it seems that what Georgi is describing corresponds to the variables in a division algebra. For example, the group C_2 has the elements:

$$\begin{bmatrix} 1 & 0 \\ 0 & 1 \end{bmatrix}, \quad \begin{bmatrix} 0 & 1 \\ 1 & 0 \end{bmatrix} \qquad (20.2)$$

At first, it might seem that we associate the set of continuous parameters (a,b) with this group as:

---

[74] Howard Georgi: Lie Algebras in Particle Physics ISBN: 978-0-7382-0233-4 Chapter 2 Lie Algebras.

$$\begin{bmatrix} a & 0 \\ 0 & a \end{bmatrix}, \quad \begin{bmatrix} 0 & b \\ b & 0 \end{bmatrix} \tag{20.3}$$

but, as we read on, we will have to change this view.

"... it is useful to parameterize the elements ... in such a way that $\alpha = 0$ corresponds to the identity element. Thus we assume ... the group elements can be described by a function of N real parameters, α_a for $a = 1$ to N, such that

$$g(\alpha)|_{\alpha=0} = e \tag{20.4}"$$

We see that we need to change the way we have parameterized the group to accommodate the "$\alpha = 0$ *corresponds to the identity element*" part of Georgi's requirements. This is easy; we have to forget the actual identity element, the real variable, and work with only the imaginary variables; we need the exponential of these imaginary elements to form the rotation matrix:

$$\exp\left(\begin{bmatrix} 0 & b \\ b & 0 \end{bmatrix}\right) = \begin{bmatrix} \cosh b & \sinh b \\ \sinh b & \cosh b \end{bmatrix} \tag{20.5}$$

We see that when $b = 0$, this becomes the identity:

$$\begin{bmatrix} 1 & 0 \\ 0 & 1 \end{bmatrix} \tag{20.6}$$

It is the nature of rotation matrices that the leading trigonometric function is unity when the angle is zero; it has to be because a rotation through zero angle always gives the identity. The other trigonometric functions are, of course, projections on to the different axes of the division algebra.

Georgi will shortly call the N real parameters mentioned above the N generators of the Lie group. Within division algebras, we might see the N generators as the $(N-1)$ different imaginary variables of a N-dimensional division algebra or we might see them as the N trigonometric functions within the rotation matrix of that algebra.

"Then if we find a representation of the group, the linear operators of the representation will be parameterized in the same way, and

$$D(\alpha)|_{\alpha=0}=1 \qquad (20.7)$$

... we can Taylor expand $D(\alpha)$, and if we are close enough, just keep the first term:

$$D(d\alpha)=1+id\alpha_a X_a +... \qquad (20.8)$$

...

$$X_a \equiv -i\frac{\partial}{\partial\alpha_a} D(\alpha)|_{\alpha=0} \qquad (20.9)$$

*The X_a for $a=1$ to N are called the **generators of the group**. ... The i is included in the definition (20.9) so that if the representation is unitary, the X_a will be hermitain operators."*

We assert that $D(\alpha)$ corresponds to a rotation matrix. The leading trigonometric function, like $\cos(\ )$, is one when the rotation angle is zero. It is the nature of trigonometric functions that they differentiate into each other (but see below (21.4)). In 3-dimensions, for example, we have:

$$\frac{\partial v_A(b,c)}{\partial b}=v_C$$
$$\frac{\partial v_A(b,c)}{\partial c}=v_B \qquad (20.10)$$

Matrices containing only these are the generators, but also the leading trigonometric function is a generator. The reader will see that we have a confusion between the Cartesian imaginary variables and the trigonometric functions that are in the rotation matrix.

Since we write our division algebras as matrices, and the off diagonal elements are all imaginary (though not all i), we have no need of the i which Georgi inserts. Because we deal with symmetric matrices in the $C_2 \times C_2 \times ...$ groups, we have no interest in hermitian matrices; symmetric matrices are hermitian matrices in a simpler notation.

"This suggests defining the representation of the group elements for finite α as

$$D(\alpha) = ... = e^{i\alpha_a X_a} \qquad (20.11)$$

... the generators form a vector space."

We put, for example in 3-dimensions[75]:

$$e^{i\alpha_a} = e^{\begin{bmatrix} 0 & b & 0 \\ 0 & 0 & b \\ b & 0 & 0 \end{bmatrix}} \qquad : \qquad e^{i\alpha_a} = e^{\begin{bmatrix} 0 & 0 & c \\ c & 0 & 0 \\ 0 & c & 0 \end{bmatrix}} \qquad (20.12)$$

We now seem to have undone the confusion between the imaginary Cartesian variables and the trigonometric functions. Georgi's generators form a vector space; our generators, together with the discarded real variable, are a division algebra; without the discarded real variable, our generators are the angular part of a division algebra – they correspond to a unit sphere in the division algebra space. Such a unit sphere is a group of infinite order.

"Now, in any particular direction, the group multiplication law is uncomplicated. There is a one parameter family of group elements of the form

$$U(\lambda) = e^{i\lambda\alpha_a X_a} \qquad (20.13)$$

and the group multiplication law is simply

$$U(\lambda_1)U(\lambda_2) = U(\lambda_1 + \lambda_2) \qquad (20.14)"$$

The group elements depend smoothly on the parameter, λ. Although Georgi does not say so, we say λ_i are 2-dimensional angles and the

---

[75] Taking the exponential of a single commutative variable leads to a rotation matrix with only one variable. The product of such matrices leads to the full rotation matrix. See Dennis Morris: Complex Numbers The Higher Dimensional forms Page 96.

$U(\lambda_i)$ are rotation matrices. Georgi uses the phrase 'particular direction' to mean rotation in a 2-dimensional plane. Georgi is unaware of higher dimensional commutative rotations such as the 3-dimensional one above, and so he includes 'particular direction'. The i in the exponential is a notational way of saying that we get rotations by taking the exponential of the imaginary variables. We continue with Georgi.

"However, if we multiply group elements generated by two different linear combinations, things are not so easy. In general

$$e^{i\alpha_a X_a} e^{i\beta_a X_b} \neq e^{i(\alpha_a + \beta_a) X_c} \qquad (20.15)$$

On the other hand, because the exponentials form a representation of the group (at least if we are close to the identity), it must be true that the product is some exponential of a generator

$$e^{i\alpha_a X_a} e^{i\beta_a X_b} = e^{i\delta_a X_c} \qquad (20.16)$$

for some δ. And because everything is smooth, we can find δ_a by expanding both sides and equating appropriate powers of α and β."

Of course (20.15) is an equality if the generators are commutative, and so Georgi is going to ignore commutative generators. Putting our words in Georgi's mouth, Georgi is saying that a product of two non-commutative rotations through, say, (b,c,d) and (f,g,h) is a rotation even though it is not a rotation through the angle, say, $(b+f,\ c+g,\ d+h)$. Of course it is a rotation. We continue with Georgi.

"When we do this, (expand both sides of (20.16)) something interesting happens. We find that it works only if the generators form an algebra under commutation (or a commutator algebra)."

By commutator algebra, Georgi means, as he later explains:

$$[X_a, X_b] = X_a X_b - X_b X_a = i f_{abc} X_c \qquad (20.17)$$

The essence of this is that the commutator of two generators gives, subject to multiplication by if_{abc}, another generator. f_{abc} is assumed to be real. This is multiplicative closure of the generators except that the multiplication is not normal matrix multiplication but a commutator operation. Within the quaternions, this is more familiar as the commutation relations:

$$\hat{i}\hat{j} = \hat{k}, \quad \hat{j}\hat{k} = \hat{i}, \quad \hat{k}\hat{i} = \hat{j}$$
$$\hat{j}\hat{i} = -\hat{k}, \quad \hat{k}\hat{j} = -\hat{i}, \quad \hat{i}\hat{k} = -\hat{j} \quad\quad (20.18)$$

*"The commutator relation (20.17) is called the Lie algebra of the group. ... Now, you might worry that if we keep expanding (20.16) beyond second order, we would need additional conditions to make sure that the group multiplication law is maintained. **The remarkable thing is that we don't. The commutator relation (20.17) is enough.**"* (Georgi's emboldenment).

Slightly before the above quotation (first line on page 47), Georgi says, *"... so that if γ and the higher terms vanish, we would restore the equality in (20.15)."* Although Georgi does not expand this statement, and perhaps he does not realise this, Georgi is effectively saying that γ and the higher order terms are a measure of the amount of non-commutativity.

Georgi goes on to describe the adjoint representation.

"The dimension of the adjoint representation is just the number of independent generators."

Thus, the dimension of the adjoint representation of $SU(2)$ is three. To us, $SU(2)$ is associated with the 4-dimensional quaternions, but of this division algebra, $SU(2)$ is the 4×4 rotation matrix which is the exponential of the three imaginary variables. We will associate the adjoint representation with the rotation matrix of our division algebras.

Between the standard Lie algebra and the division algebra Lie algebra, there is the confusion between the Cartesian imaginary variables of an algebra, which can appear as separate matrices, and the trigonometric functions of the algebra which must appear together in a rotation matrix. Georgi's generators are the trigonometric functions within a rotation matrix, but he also presents them as separate variables. The two presentations are correlated in that the rotation matrix is derived from a finite group by taking the exponential of the imaginary variables. Georgi's generators have commutation relations between them, as do the individual imaginary variables and as would the individual trigonometric functions if we could separate them from the rotation matrix.

So, we assert that the standard Lie groups are 'spherical surfaces' in $\{\mathbb{R}^n, \mathbb{C}^n, \mathbb{H}^n, \mathbb{O}^n\}$ types of space and that division algebra Lie groups are 'spherical surfaces' in division algebra spaces. The previous sentence sums up the difference between standard Lie algebra and division algebra Lie algebra.

The standard generators are 'really' imaginary variables whose exponentials are 2-dimensional rotations. Division algebra generators are the imaginary variables in the zero trace algebraic matrix form whose exponential is the n-dimensional rotation matrix. Division algebra rotations are, in general, higher dimensional rotations rather than 2-dimensional rotations.

What would a 'spherical surface' look like? Well except for the absent radial variable, a 'spherical surface' is what we think of as empty space. We find the whole of the spatio-temporal part of special relativity within the 2-dimensional hyperbolic rotation matrix; all the radial component does is move this away from the origin.

The Killing form and the quaternion inner product:
In quest of an inner product over the linear space that is the generators, Lie algebraists invent the Killing form.

"We would like to have a convenient scalar product on the linear space of the generators in the adjoint representation ... to turn it into a vector space. A good one is the trace in the adjoint representation

$$Trace\left(T_aT_b\right) \qquad (20.19)\text{''}$$

Since we identify the generators with the trigonometric functions, this would be multiplying a quaternion rotation matrix with one non-zero variable by a similar matrix with a different non-zero variable; this is artificial reduction of the quaternions to 2-dimensional algebras, and so we will not walk that path.

In the commutative algebras, the trace of a product of two suitable n-dimensional rotations is just n times the inner product (the leading trigonometric function of the angle between the vectors.) We can see that the Killing form is just n times the inner product, but this inner product is based upon a commutative rotation and the conception that the angle between two vectors in, say, 4-dimensional space is a 2-dimensional angle. Within the quaternions, because they are non-commutative, as shown above, this leading trigonometric function is not a direct measure of the angle between two vectors. We reject the Killing form. Of course, we must replace it with something better. To do this, we will first have to look at non-commutative rotation in more detail, but, before we do that, we will briefly consider the traditional quaternion inner product.

Traditionally, the inner product of two quaternions in Cartesian form is defined as:

$$\left\langle \begin{matrix} a & e \\ \hat{i}b & \hat{i}f \\ jc & jg \\ kd & kh \end{matrix} \right\rangle = \frac{ae + bf + cg + dh}{\sqrt{a^2 + b^2 + c^2 + d^2}\sqrt{e^2 + f^2 + g^2 + h^2}} \qquad (20.20)$$

Based on the belief that any two quaternions are separated by a 2-dimensional angle, this inner product is traditionally set equal to the cosine of a 2-dimensional angle:

$$\cos\left(\theta\right) = \frac{ae + bf + cg + dh}{\sqrt{a^2 + b^2 + c^2 + d^2}\sqrt{e^2 + f^2 + g^2 + h^2}} \qquad (20.21)$$

Quaternions are not separated by a 2-dimensional angle; quaternions are separated by a 4-dimensional angle, and so at least the LHS of (20.21) is incorrect. The reader might think all we do is change the argument of the cosine giving:

$$\cos\left(\sqrt{(\alpha-\eta)^2+(\beta-\kappa)^2+(\chi-\lambda)^2}\right)$$
$$\overset{?}{=}\frac{ae+bf+cg+dh}{\sqrt{a^2+b^2+c^2+d^2}\sqrt{e^2+f^2+g^2+h^2}}$$

(20.22)

This does not work; the two sides are just not equal. Since the expression $\sqrt{(\alpha-\eta)^2+(\beta-\kappa)^2+(\chi-\lambda)^2}$ does not generally appear anywhere in the product of two polar quaternions, we might have expected this would not be within the inner product. We need a way of defining a non-commutative inner product.

The quaternion inner product:
We have the two quaternion rotation matrices:

$$A=\begin{bmatrix} \cos(\lambda) & \frac{b}{\lambda}\sin(\lambda) & \frac{c}{\lambda}\sin(\lambda) & \frac{d}{\lambda}\sin(\lambda) \\ -\frac{b}{\lambda}\sin(\lambda) & \cos(\lambda) & -\frac{d}{\lambda}\sin(\lambda) & \frac{c}{\lambda}\sin(\lambda) \\ -\frac{c}{\lambda}\sin(\lambda) & \frac{d}{\lambda}\sin(\lambda) & \cos(\lambda) & -\frac{b}{\lambda}\sin(\lambda) \\ -\frac{d}{\lambda}\sin(\lambda) & -\frac{c}{\lambda}\sin(\lambda) & \frac{b}{\lambda}\sin(\lambda) & \cos(\lambda) \end{bmatrix}$$ (20.23)
$$\lambda=\sqrt{b^2+c^2+d^2}$$

and:

$$C = \begin{bmatrix} \cos(\beta) & \dfrac{f}{\beta}\sin(\beta) & \dfrac{g}{\beta}\sin(\beta) & \dfrac{h}{\beta}\sin(\beta) \\[2ex] -\dfrac{f}{\beta}\sin(\beta) & \cos(\beta) & -\dfrac{h}{\beta}\sin & \dfrac{g}{\beta}\sin(\beta) \\[2ex] -\dfrac{g}{\beta}\sin(\beta) & \dfrac{h}{\beta}\sin & \cos(\beta) & -\dfrac{f}{\beta}\sin(\beta) \\[2ex] -\dfrac{h}{\beta}\sin & -\dfrac{g}{\beta}\sin(\beta) & \dfrac{f}{\beta}\sin(\beta) & \cos(\beta) \end{bmatrix} \quad (20.24)$$

$$\beta = \sqrt{f^2 + g^2 + h^2}$$

We are in search of an inner product, and so we will take the conjugate of the second matrix, C^*, rather than the matrix as it stands. The conjugate of a rotation matrix is just rotation in the reverse direction. By multiplying these together, we get a measure of the angle between the two quaternions $(b - f, \ c - g, \ d - h)$.

The inner product appears on the leading diagonal of the product, $C * A$. The order of multiplication does not affect the element of the leading diagonal, and so the inner product is unique. The leading element of the product is:

$$C * A_{[1,1]} = \cos\left(\sqrt{b^2 + c^2 + d^2}\right)\cos\left(\sqrt{f^2 + g^2 + h^2}\right)$$

$$+ \frac{(bf + cg + dh)\sin\left(\sqrt{b^2 + c^2 + d^2}\right)\sin\left(\sqrt{f^2 + g^2 + h^2}\right)}{\sqrt{b^2 + c^2 + d^2}\sqrt{f^2 + g^2 + h^2}}$$

$$(20.25)$$

The two rotation matrices above, (20.23) and (20.24) are equal to two normalised Cartesian quaternion matrices:

$$A = \frac{1}{\sqrt{t^2 + x^2 + y^2 + z^2}} \begin{bmatrix} t & x & y & z \\ -x & t & -z & y \\ -y & z & t & -x \\ -z & -y & x & t \end{bmatrix} \quad (20.26)$$

$$C = \frac{1}{\sqrt{p^2+q^2+r^2+s^2}} \begin{bmatrix} p & q & r & s \\ -q & p & -s & r \\ -r & s & p & -q \\ -s & -r & q & p \end{bmatrix}$$

(20.27)

Calculating the product of the Cartesian matrices $C*A$ (note that C is conjugated) gives:

$$C*A_{[1,1]} = \frac{pt+qx+ry+sz}{\sqrt{t^2+x^2+y^2+z^2}\sqrt{p^2+q^2+r^2+s^2}}$$

(20.28)

We can put this equal to the polar form above, (20.25).

$$\cos\left(\sqrt{b^2+c^2+d^2}\right)\cos\left(\sqrt{f^2+g^2+h^2}\right)$$
$$+ \frac{(bf+cg+dh)\sin\left(\sqrt{b^2+c^2+d^2}\right)\sin\left(\sqrt{f^2+g^2+h^2}\right)}{\sqrt{b^2+c^2+d^2}\sqrt{f^2+g^2+h^2}}$$

(20.29)

$$= \frac{pt+qx+ry+sz}{\sqrt{t^2+x^2+y^2+z^2}\sqrt{p^2+q^2+r^2+s^2}}$$

This is our non-commutative quaternion inner product.

We note that the trigonometric expression in (20.29) also occurs in the measure of non-commutativity, Ω, given above, (19.14). If we want, we can substitute the Cartesian inner product into our measure of non-commutativity. We prefer this over the Killing form which we think is meaningless.

Geodesics and non-commutativity:

Consider the 'spherical surface' that is the Lie group of a non-commutative division algebra. Consider an object moving from point X within this 'surface' to point Y within this 'surface'. The object might travel the first part of its journey from point X in one direction and then change direction to travel the remainder of its journey to point Y. Both of these sub-journeys are rotations through particular angles. Suppose that the two particular angles point in the same direction. This means that one angle is just a multiple of the other angle. Let the angles be (b,c,d) and (nb, nc, cd); the amount of non-commutativity of the two rotations is:

$$\cos\left(\sqrt{(b+nb)^2 + (c+nc)^2 + (d+nd)^2}\right) -$$
$$\cos\left(\sqrt{b^2 + c^2 + d^2}\right)\cos\left(n\sqrt{b^2 + c^2 + d^2}\right) + \quad (20.30)$$
$$\frac{n\left(b^2 + c^2 + d^2\right)\sin\left(\sqrt{b^2 + c^2 + d^2}\right)\sin\left(n\sqrt{b^2 + c^2 + d^2}\right)}{n\sqrt{b^2 + c^2 + d^2}\sqrt{b^2 + c^2 + d^2}} = \Omega$$

Simplifying (20.30):

$$\cos\left((n+1)\sqrt{b^2 + c^2 + d^2}\right) - \cos\left((n+1)\sqrt{b^2 + c^2 + d^2}\right) = 0 = \Omega \quad (20.31)$$

We see, as we might intuitively have expected, that two rotations in the same direction are commutative. Clearly, this would not be the case if the rotations were not in the same direction. Of course, what we have done with two rotations, we could have done with a billion smaller rotations. Intuitively, we can take it that successive rotations in the same direction are a 'straight line' (great circle) in the 'spherical surface' space. We are therefore able to define a geodesic within the Lie group that is the 'spherical surface' by the measure of non-commutativity. When there is zero non-commutativity between two successive rotations, those two rotations lie upon a geodesic within the Lie group.

We think that, if non-commutativity is associated with deviation from a geodesic, then non-commutativity is associated with a force.

The nature of force:

If I move a magnet close to a small ferric object, at some point, that object will 'jump' toward the magnet. Some while ago, Isaac Newton used to speak of such a jump as 'action at a distance' when describing such phenomenon. Michael Faraday did not like 'action at a distance' and replaced it with the concept of a force field around the magnet. The mathematics of the force field works well, but the concept is no better founded than the 'action at a distance' which it replaced.

At various places above, we have seen different ideas of what a force might be:

1) The difference between the distance function of quaternion space and the distance function of the observer's space-time might be the nature of force.
2) The correlation between deviation from a geodesic in a division algebra Lie group space and the amount of non-commutativity might be the nature of force.
3) The different relative scaling of the six A_3 spaces might be the nature of force.

We might speculate that these different views of force might correspond to different types of force. Perhaps the difference in distance functions of quaternion space and space-time is the electro-magnetic force; perhaps the correlation between deviation from a geodesic and non-commutativity in quaternion space is the weak force; perhaps the different scaling of the six A_3 spaces is the gravitational force.

As your author writes, it is probably fair to say that we still have no proper understanding of the nature of force.

Aside: Particle physicists see force as being a consequence of the mediation of virtual particles between two bodies.

Chapter 21

Differentiation of the Quaternion Trig. Functions

When dealing with the trigonometric functions of commutative division algebras, we habitually differentiate a single trigonometric function without regard to it being within a rotation matrix[76]. We differentiate non-commutatively within the quaternion algebra. Non-commutative differentiation does not allow us to deal with a trigonometric function on its own because non-commutativity 'needs matrices[77]' and these matrices must act on matrices.

Such non-commutative differentiation leads to the E differential and the B differential defined as:

$$E^{Diff} = \frac{1}{2}\left(d_L + d_R\right)$$
$$B^{Diff} = \frac{1}{2}\left(d_L - d_R\right)$$

(21.1)

We act upon the quaternion rotation matrix with the $SU(2)$ differential operator (or we differentiate properly though cumbersomely) to both the left and to the right to form the left differential, d_L, and the right differential, d_R.

---

[76] We do so differentiate trigonometric functions and have done so since we were in our nappies, but is it correct? Do trigonometric functions exist outside of a rotation matrix? Are we mistreating them when we take them from where they belong?

[77] A definite lack of precision here.

227

$$d_L = PROD \begin{pmatrix} \begin{bmatrix} \partial a & -\partial b & -\partial c & -\partial d \\ \partial b & \partial a & \partial d & -\partial c \\ \partial c & -\partial d & \partial a & \partial b \\ \partial d & \partial c & -\partial b & \partial a \end{bmatrix} \\ \begin{bmatrix} \cos(\lambda) & \dfrac{b}{\lambda}\sin(\lambda) & \dfrac{c}{\lambda}\sin(\lambda) & \dfrac{d}{\lambda}\sin(\lambda) \\ -\dfrac{b}{\lambda}\sin(\lambda) & \cos(\lambda) & -\dfrac{d}{\lambda}\sin(\lambda) & \dfrac{c}{\lambda}\sin(\lambda) \\ -\dfrac{c}{\lambda}\sin(\lambda) & \dfrac{d}{\lambda}\sin(\lambda) & \cos(\lambda) & -\dfrac{b}{\lambda}\sin(\lambda) \\ -\dfrac{d}{\lambda}\sin(\lambda) & -\dfrac{c}{\lambda}\sin(\lambda) & \dfrac{b}{\lambda}\sin(\lambda) & \cos(\lambda) \end{bmatrix} \\ \lambda = \sqrt{b^2 + c^2 + d^2} \end{pmatrix} \quad (21.2)$$

$$d_R = PROD \begin{pmatrix} \begin{bmatrix} \cos(\lambda) & \dfrac{b}{\lambda}\sin(\lambda) & \dfrac{c}{\lambda}\sin(\lambda) & \dfrac{d}{\lambda}\sin(\lambda) \\ -\dfrac{b}{\lambda}\sin(\lambda) & \cos(\lambda) & -\dfrac{d}{\lambda}\sin(\lambda) & \dfrac{c}{\lambda}\sin(\lambda) \\ -\dfrac{c}{\lambda}\sin(\lambda) & \dfrac{d}{\lambda}\sin(\lambda) & \cos(\lambda) & -\dfrac{b}{\lambda}\sin(\lambda) \\ -\dfrac{d}{\lambda}\sin(\lambda) & -\dfrac{c}{\lambda}\sin(\lambda) & \dfrac{b}{\lambda}\sin(\lambda) & \cos(\lambda) \end{bmatrix} \\ \begin{bmatrix} \partial a & -\partial b & -\partial c & -\partial d \\ \partial b & \partial a & \partial d & -\partial c \\ \partial c & -\partial d & \partial a & \partial b \\ \partial d & \partial c & -\partial b & \partial a \end{bmatrix} \\ \lambda = \sqrt{b^2 + c^2 + d^2} \end{pmatrix} \quad (21.3)$$

The resulting differentials are:

$$E_{[1,1]}^{Diff} = \frac{\partial}{\partial a}\cos(\lambda) + \frac{\partial}{\partial b}\left(\frac{b}{\lambda}\sin(\lambda)\right) + \frac{\partial}{\partial c}\left(\frac{c}{\lambda}\sin(\lambda)\right) + \frac{\partial}{\partial d}\left(\frac{d}{\lambda}\sin(\lambda)\right)$$

$$= 0 + \frac{b^2\lambda\cos(\lambda) + c^2\sin(\lambda) + d^2\sin(\lambda)}{\lambda^3}$$

$$+ \frac{c^2\lambda\cos(\lambda) + b^2\sin(\lambda) + d^2\sin(\lambda)}{\lambda^3}$$

$$+ \frac{d^2\lambda\cos(\lambda) + b^2\sin(\lambda) + c^2\sin(\lambda)}{\lambda^3}$$

$$(21.4)$$

This simplifies to:

$$E_{[1,1]}^{Diff} = \cos(\lambda) + 2\frac{\sin(\lambda)}{\lambda} \qquad (21.5)$$

With:

$$E_{[1,2]}^{Diff} = \frac{b}{\lambda}\sin(\lambda) \qquad E_{[1,3]}^{Diff} = \frac{c}{\lambda}\sin(\lambda) \qquad E_{[1,4]}^{Diff} = \frac{d}{\lambda}\sin(\lambda)$$

$$(21.6)$$

And:

$$B_{[1,1]}^{Diff} = B_{[1,2]}^{Diff} = B_{[1,3]}^{Diff} = B_{[1,4]}^{Diff} = 0 \qquad (21.7)$$

We see that, with the exception of the cosine on the leading diagonal, the rotation matrix has differentiated into itself. (21.4) is deceptive; we note that if two of the variables are zero, the leading diagonal becomes:

$$E_{[1,1]}^{Diff} = \frac{\partial}{\partial a}\cos(\lambda) + \frac{\partial}{\partial b}\left(\frac{b}{\lambda}\sin(\lambda)\right) + \frac{\partial}{\partial c}\left(\frac{c}{\lambda}\sin(\lambda)\right) + \frac{\partial}{\partial d}\left(\frac{d}{\lambda}\sin(\lambda)\right)$$

$$= \frac{\partial}{\partial b}\left(\frac{b}{b}\sin(b)\right) = \cos(b)$$

$$(21.8)$$

The Expanding Universe (perhaps)

In this chapter, we present an explanation of why the universe is expanding. Basically, we think this is because the 2-dimensional space-time 'plane' is the inside of a hyperbola and the 'arms' of the hyperbola get further apart with increasing real variable. We associate the real axis with the age of the universe and the 'arms' of the hyperbola with the edges of the observable universe; thus, as we progress along the real axis (we age) the edges of the universe get further apart and the space between those edges gets stretched. There is a problem with our explanation in that there is no sign of the accelerating expansion that cosmologists seem to have detected. We do, however, have an inflationary start to the universe without any need of phase transitions or the like. It is for the reader to judge the merits of this explanation.

Space-time vectors:
A vector in 2-dimensional space-time is of the form:

$$\begin{bmatrix} e^a & 0 \\ 0 & e^a \end{bmatrix} \begin{bmatrix} \cosh \chi & \sinh \chi \\ \sinh \chi & \cosh \chi \end{bmatrix} \tag{22.1}$$

For a given value of a, the totality of these vectors are points which, when taken all together, are a hyperbola about the real (horizontal) axis that cuts the real axis at e^a. Of course, e^a is always greater than zero and, when $a = 0$, $e^a = 1$. As a increases, the space spanned by the vectors becomes a succession of hyperbolae 'inside' the original hyperbola; each of these 'inside hyperbolae' cuts the real axis at the new value of e^a. By this means, the 'inside' of the hyperbola below is 'filled in solidly'. Thus, we see that the shape of 2-dimensional space-time is a 'solid' hyperbola.

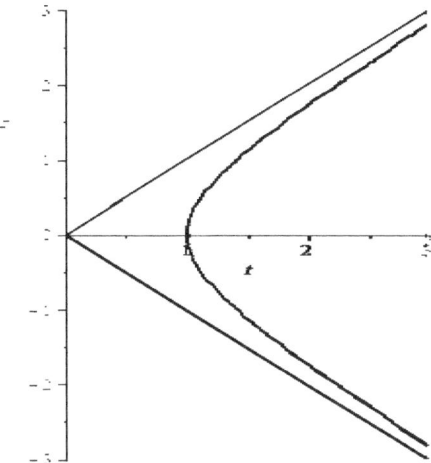

The units of the above graph are based upon the limiting velocity of the universe being unity, and the graph is squashed in the vertical direction.

One of the fundamental types of vector is the displacement vector. The 2-dimensional space-time displacement vector is therefore a point within the hyperbola shown above. Such a displacement vector is a position in space-time; that is a particular spatial position and a particular time. The time is the position on the horizontal axis. The spatial position is the position on the vertical axis.

We identify a with the time component of the 2-dimensional space-time vector (the age of the universe). We identify χ with the spatial component of the 2-dimensional space-time vector. We identify the points along the 'outside' curve, that is the hyperbola that cuts the real axis at unity, with the spatial edge of the universe. An observer views the universe as if he were sitting upon the real axis moving through time (aging) towards the right (increasing real values of a). We see immediately that, as the observer ages, the spatial extent of the universe increases. We think this is the expanding universe. The rate of expansion is the gradient of the hyperbola. We see that age zero is unity on the real axis. Unity is the least possible value of the $\cosh(\ )$ function which is the time component.

If we look back along the real axis to the point when the age of the universe, a, was zero (that is unity on the real axis), we see that the universe was spatially expanding at an infinite rate at age zero. This explains the homogeneity of the universe and replaces the concept of inflation in that regard.

Observations indicate that the rate of expansion is increasing as the universe ages whereas the gradient of the hyperbolae which we identified with the rate of expansion is decreasing as the universe ages. Perhaps this observed acceleration of expansion is something to do with gravity losing its grip as space expands or perhaps it is something to do with the changing limiting velocity of the universe or perhaps we have misunderstood the expanding universe part of space theory. Your author's personal view is that the acceleration is an optical illusion. None-the-less, at this moment in time, we cannot claim this is a completely acceptable explanation of the expanding universe.

Chapter 23

8-dimensional Space

We have derived our 4-dimensional space-time by aggregating together the 4-dimensional A_3 algebras from within the $C_2 \times C_2$ finite group. There are 8-dimensional algebras in the $C_2 \times C_2 \times C_2$ group that could be aggregated together, and so we would expect to see an 8-dimensional space-time. Similarly, there are higher dimensional $C_2 \times C_2 \times ...$ groups which should produce higher dimensional space-times. Why do we not see these higher dimensional space-times?

Remarkably, the higher dimensional $C_2 \times C_2 \times ...$ spaces fold into a 4-dimensional form. This is not an imagined 'folding' similar to the 'rolling up' of higher dimensions postulated in string theory to explain the absence of observed higher spatial dimensions. This folding is within the mathematics.

The 8-dimensional algebras each have seven 4-dimensional sub-algebras. These 4-dimensional sub-algebras have 4-dimensional trigonometric functions, and so the 8-dimensional trigonometric functions must be of a forms that reduce to the 4-dimensional trigonometric functions when four of the variables are zero. Furthermore, they must do this for each of the seven combinations of four variables that correspond to the seven 4-dimensional sub-algebras. They must also reduce to the 2-dimensional trigonometric functions for each of the seven 2-dimensional sub-algebras. Calculation shows that the 8-dimensional trigonometric functions achieve this by pairing together the eight variables into four pairs. It is not known if this could be achieved without such pairing of variables, but this does not matter. What matters is that the 8-dimensional trigonometric functions do pair variables together.

Similarly, the 16-dimensional trigonometric functions pair together 8-dimensional pairs of 4-dimensional variables. There are thirty-five 4-dimensional sub-algebras within a 16-dimensional algebra and fifteen 8-dimensional sub-algebras also.

Not only do the 8-dimensional trigonometric functions need to accommodate the 4-dimensional trigonometric functions but the 8-dimensional distance function (the determinant) must accommodate the 4-dimensional distance functions. This is also achieved by pairing variables together. Remarkable functions!

8-dimensional trigonometric functions:

With the $\{a,e\}$ variables set to zero[78], taking the exponential of a particular (double quaternion) 8-dimensional algebra matrix form produces a rotation matrix with trigonometric function like:

$$DQ_{[1,1]}^{NC} = \frac{1}{2}\left(\begin{array}{c} \cos\left(\sqrt{(b+f)^2 + (c+g)^2 + (d+h)^2}\right) \\ + \cos\left(\sqrt{(b-f)^2 + (c-g)^2 + (d-h)^2}\right) \end{array} \right) \quad (23.1)$$

The other results are of the form:

$$DQ_{[1,2]}^{NC} = \frac{1}{2}\left(\begin{array}{c} \dfrac{(b+f)\sin\left(\sqrt{(b+f)^2 + (c+g)^2 + (d+h)^2}\right)}{\sqrt{(b+f)^2 + (c+g)^2 + (d+h)^2}} \\ + \dfrac{(b-f)\sin\left(\sqrt{(b-f)^2 + (c-g)^2 + (d-h)^2}\right)}{\sqrt{(b-f)^2 + (c-g)^2 + (d-h)^2}} \end{array} \right) \quad (23.2)$$

The complete 8-dimensional rotation matrix has trigonometric functions like:

---

[78] The e variable is commutative.

$$DQ^{ROT}_{[1,1]} = \cosh(e)\,DQ^{NC}_{[1,1]} + \sinh(e)\,DQ^{NC}_{[1,5]}$$

$$DQ^{ROT}_{[1,2]} = \cosh(e)\,DQ^{NC}_{[1,2]} + \sinh(e)\,DQ^{NC}_{[1,6]}$$

(23.3)

This pairing together of variables gives the 8-dimensional spaces a 4-dimensional characteristic. We think this is why we do not observe a space-time of more dimensions than four. However, this area of the mathematics is poorly understood.

Chapter 24

Concluding Remarks

In earlier chapters, we have laid out a theory of empty space. The theory is still in its infancy; it is hardly known in the physics community; it is still incomplete, and there are parts of it that might be misinterpreted. None-the-less, we think it does have promise.

Of particular interest is the concept of mathematics being divided into quantum mathematics (division algebras) and classical mathematics (tensors etc.) that is derived by aggregating together division algebras. We do not properly understand this aggregation of division algebras, but the division of mathematics into quantum and classical parts is a remarkable match with the physical world.

In forming our theory of space, we have postulated no more than the existence of real numbers and finite groups. Arguably, the finite groups we need occur as relations between sets of real numbers and so are within the real numbers anyway. Further, philosophers might say we can get the real numbers from no more than the number one, and so perhaps we postulate no more than the existence of the number one. We do not need to postulate wiggly strings or 10-dimensional space. This very minimal postulation is a most desirable property for a physical theory, and it is hard to think of any more minimal postulate than only the real numbers and the finite groups.

The mathematics we have used is simple stuff being no more than matrices. Simplicity is a most desirable property for a physical theory. It is starting to look like the universe is no more than different types of numbers interacting together somehow.

236

Successes:

We have produced an intrinsically quantitised form of electromagnetism (with anti-matter); this is not QED, but it might become so. By aggregation of algebras, we have produced the electromagnetic tensor.

By aggregation, we have produced the distance function and 2-dimensional rotation matrices of the space-time in which we sit. We have been able to construct our observed 4-dimensional space-time and reproduce the field equations of general relativity.

We have a unification of gravitation and classical electromagnetism.

We have an explanation of where the physical constants come from and of their nature. We have produced an alternative form of Lie group theory. We have produced a 4-dimensional version of the Pauli matrices. We have produced the momentum operator together with the physical constant $\hbar$.

We have produced the spatial-temporal part of the theory of special relativity completely (we have no mass or electric charge). We might have the expanding universe with an inflationary beginning. We have measured non-commutativity and have associated it with deviation from a geodesic.

Perhaps most important of all successes is an aspect of the mathematics that we have only mentioned in this book. The rotation matrices of division algebras are normalised spinors.

Speculative bits:

There are three pairs of A_3 spaces within the space-time in which we sit. If these algebras are associated with the masses of the elementary particles, then we would have three generations of particles; more accurately, one generation of particles in which each particle has three different masses.

We found electromagnetism and $SU(2)$ in the 4-dimensional $C_2 \times C_2 \times \ldots$ spaces. Will we find the strong force in the 8-dimensional spaces? If we do, we might well find super-symmetric forces in the 16-dimensional spaces. If this does turn out to be so, then we will have the ability to calculate at arbitrarily high energies. At present, we do not really know how an 8-dimensional space will manifest itself to beings who view the world with 4-dimensional electromagnetism.

Failures:

We seem to get an expanding universe whose expansion is slowing rather than accelerating.

We do not yet understand the higher dimensional algebras properly .

Chapter 25

Appendix: Properties of the $C_2 \times C_2 \times ...$ Algebras

The basic algebraic form:

The basic algebraic matrix form (the one without scaling parameters) of the $C_2 \times C_2 \times ...$ algebras is symmetrical in various ways. For example:

$$C_2 \times C_2 \cong \begin{bmatrix} a & b & c & d \\ b & a & d & c \\ c & d & a & b \\ d & c & b & a \end{bmatrix} \qquad (25.1)$$

We have that the first column matches the top row, the last column is the reverse of the top row, and the bottom row is the reverse of the top row. We have that the third row is the reverse of the second row and that the third column is the reverse of the second column. These are consequences of the reflective symmetries in both the leading diagonal and the opposing diagonal. These two reflective symmetries are general to all $C_2 \times C_2 \times ...$ algebras. Other division algebras do not have these reflective symmetries; for example:

$$C_4 \cong \begin{bmatrix} a & b & c & d \\ d & a & b & c \\ c & d & a & b \\ b & c & d & a \end{bmatrix} \cong \begin{bmatrix} a & b & c & d \\ c & a & d & b \\ b & d & a & c \\ d & c & b & a \end{bmatrix} \qquad (25.2)$$

We see that matrix form (25.1) is comprised of 2×2 blocks of the form:

$$C_2 \cong \begin{bmatrix} a & b \\ b & a \end{bmatrix} \qquad (25.3)$$

This too is a general phenomenon of to all $C_2 \times C_2 \times \ldots$ algebras. Further the 8×8 algebras are also 4×4 blocks of the form (25.1) and the 16×16 algebras are also 8×8 blocks (and obviously 4×4 blocks and 2×2 blocks) and so it proceeds to higher order algebras *ad infinitum*.

There is another kind of symmetry which we might call 'an order symmetry' within the $C_2 \times C_2 \times \ldots$ algebras. It is common to all division algebras that a particular variable within an algebra is distributed throughout the matrix form in such a way that it occurs once and only once in every row and in every column. This must be the case because the algebras are based upon permutation matrices – one for each variable. However, in the case of the $C_2 \times C_2 \times \ldots$ algebras, there is the extra property that the 'order' of the variables dictates the matrix form. We take the order of the variables to be the order in which they occur from left to right along the top row of the matrix form. We have:

$$
\begin{bmatrix}
A_{[1,1]} & A_{[1,2]} & A_{[1,3]} & A_{[1,4]} \\
A_{[1,2]} & & & \\
A_{[1,3]} & & & \\
A_{[1,4]} & & &
\end{bmatrix}
\tag{25.4}
$$

The order of the variable is the right-most number in the subscript. Since this order is the same from top to bottom in the left-most column, we can take the left-most column to define the order if we choose. Starting from the left and moving across the matrix and from the second row and moving down the matrix, we can fill in the elements of the matrix based on the rule:

Rule: If the variable $A_{[1,n]}$ does not already occur in the column (above) or the row, then that element is that variable provided n is the lowest order variable of all such variables that do not occur elsewhere in the row or column (the left-most such variable within the top row). For example; starting with:

$$\begin{bmatrix} A_{[1,1]} & A_{[1,2]} & A_{[1,3]} & A_{[1,4]} \\ & & & \\ & & & \\ & & & \end{bmatrix} \qquad (25.5)$$

We calculate the left-most element of the second row as:

Element $[2,1]$ cannot be $A_{[1,1]}$ because there is a $A_{[1,1]}$ in this column already. There is no other element in this column or in this row, and so Element $[2,1]$ might be either $\{A_{[1,2]}, A_{[1,3]}, A_{[1,4]}\}$. We choose the lowest in the order, which is $A_{[1,2]}$.

$$\begin{bmatrix} A_{[1,1]} & A_{[1,2]} & A_{[1,3]} & A_{[1,4]} \\ A_{[1,2]} & & & \\ & & & \\ & & & \end{bmatrix} \qquad (25.6)$$

Element $[2,2]$ cannot be $A_{[1,2]}$ because there is a $A_{[1,2]}$ in this row already (we just put it there) and there is an $A_{[1,2]}$ in this column already; thus, Element $[2,2]$ might be either $\{A_{[1,1]}, A_{[1,3]}, A_{[1,4]}\}$. We choose the lowest in the order, which is $A_{[1,1]}$.

$$\begin{bmatrix} A_{[1,1]} & A_{[1,2]} & A_{[1,3]} & A_{[1,4]} \\ A_{[1,2]} & A_{[1,1]} & & \\ & & & \\ & & & \end{bmatrix} \qquad (25.7)$$

Element $[2,3]$ cannot be $A_{[1,2]}$ because there is a $A_{[1,2]}$ in this row already (we put it there) and it cannot be $A_{[1,1]}$ because there is an $A_{[1,1]}$ in this row already (we just put it there) nor can it be $A_{[1,3]}$ because there is a $A_{[1,3]}$ in this column already (the top row); thus,

Element $[2,3]$ might be only $\left\{A_{[1,4]}\right\}$. The essence is that, when we have a choice of which variable to place as a particular element of the matrix, we must choose the variable of lowest order from the possible choices. The 'order symmetry' is that the order of the variables dictates which of the possible choices is to be taken. To choose with disregard to the order would lead to a different algebra. Only the $C_2 \times C_2 \times \ldots$ algebras have this 'order symmetry' – if another algebra had this 'order symmetry', then that other algebra would be the same algebra as the $C_2 \times C_2 \times \ldots$ algebra.

This 'order symmetry' facilitates writing a short computer program that will generate the $C_2 \times C_2 \times \ldots$ algebraic matrices for any order.

There is another property of the $C_2 \times C_2 \times \ldots$ algebras. They can be written in only one way (one matrix form). We see above, (25.2), that the C_4 algebra has different matrix forms (three depending upon which variable forms the C_2 sub-algebra with the identity variable). This does not happen with the $C_2 \times C_2 \times \ldots$ algebras because every imaginary element forms a C_2 sub-algebra with the identity variable. In this sense (all the imaginary variables are of equal status), the $C_2 \times C_2 \times \ldots$ algebras are more symmetric (socialist if you prefer) that any other algebras.

The reflective symmetry across the leading diagonal of the matrix means that every variable is symmetrically distributed within the matrix. We refer to these as symmetric imaginary variables. Symmetric imaginary variables are square roots of plus one. Because of the $2^n \times 2^n$ block distribution within the matrix, setting any of the scaling parameters (not yet introduced) to minus one will result in only symmetric and anti-symmetric imaginary variables. Anti-symmetric imaginary variables are square roots of minus one. Thus it is that the $C_2 \times C_2 \times \ldots$ algebras are algebras in which the imaginary variables are all square roots of ± 1. Effectively, a reflective symmetry across the leading diagonal is the same thing as saying all the variables are square roots of plus unity and a reflective anti-symmetry across the leading diagonal is equivalent to saying that all the imaginary variables are square roots of minus unity. We mention that the eigenvalues of these symmetric matrices are always real and that these symmetric matrices always have an orthogonal set of eigenvectors – symmetric matrices are like that.

The number of quadratic elimination equations:

It is the quadratic parameter elimination equations which lead to the non-commutative $C_2 \times C_2 \times ...$ algebras. The non-commutative 4-dimensional algebras, $C_2 \times C_2$, derive from only one quadratic elimination equation, $P_{4,1}$.

When we move from 4-dimensions into 8-dimensions, we can go in two directions. We can either take the positive root of the $P_{4,1}$ elimination equation or we can take the negative root of the $P_{4,1}$ elimination equation. The two choices will lead to two different 8×8 algebraic matrix forms and two sets of 512 separate 8-dimensional algebras. Similar things happen as we move from the 8-dimensional algebras to the 16-dimensional algebras and so on into the higher orders of algebras.

The number of new quadratic potential scaling parameter elimination equations seems to increase with the order of the algebra. We have:

Algebra	New Quadratics	
C_2	0	
$C_2 \times C_2$	1	
$C_2 \times C_2 \times C_2$	2	(25.8)
$C_2 \times C_2 \times C_2 \times C_2$	3	
$C_2 \times C_2 \times C_2 \times C_2 \times C_2$	5	
$C_2 \times C_2 \times C_2 \times C_2 \times C_2 \times C_2$	5	

We notice that the rate of increase of the number of quadratic elimination equations seems to be slowing. Why we get these numbers of quadratics is not understood and whether we will ever reach a point where no more are added is unknown. It seems that we will have both commutative and non-commutative algebras at all orders.

Other Books by the Same Author

The Naked Spinor – a Rewrite of Clifford Algebra

Spinors exist in Clifford algebras. In this book, we explore the nature of spinors. This book is an excellent introduction to Clifford algebra.

Complex Numbers The Higher Dimensional Forms – Spinor Algebra

In this book, we explore the higher dimensional forms of complex numbers. These higher dimensional forms are connected very closely to spinors.

Upon General Relativity

In this book, we see how 4-dimensional space-time, gravity, and electromagnetism emerge from the spinor algebras. This is an excellent and easy-paced introduction to general relativity.

From Where Comes the Universe

This is a guide for the lay-person to the physics of empty space.

Empty Space is Amazing Stuff – The Special Theory of Relativity

This book deduces the theory of special relativity from the finite groups. It gives a unique insight into the nature of the 2-dimensional space-time of special relativity.

The Nuts and Bolts of Quantum Mechanics

This is a gentle introduction to quantum mechanics for undergraduates.

Quaternions

This book pulls together the often separate properties of the quaternions. Non-commutative differentiation is covered as is non-commutative rotation and non-commutative inner products along with the quaternion trigonometric functions.

The Uniqueness of our Space-time

This book reports the finding that the only two geometric spaces within the finite groups are the two spaces that together form our universe. This is a startling finding. The nature of geometric space is explained alongside the nature of division algebra space, spinor space. This book is a catalogue of the higher dimensional complex numbers up to dimension fifteen.

Lie Groups and Lie Algebras

This book presents Lie theory from a diametrically different perspective to the usual presentation. This makes the subject much more intuitively obvious and easier to learn. Included is perhaps the clearest and simplest presentation of the true nature of the Lie group $SU(2)$ ever presented.

The Physics of Empty Space

This book presents a comprehensive understanding of empty space. The presence of 2-dimensional rotations in our 4-dimensional space-time is explained. Also included is a very gentle introduction to non-commutative differentiation. Classical electromagetism is deduced from the quaternions.

The Electron

This book presents the quantum field theory view of the electron and the neutrino. This view is radically different from the classical view of the electron presented in most schools and colleges. This book gives a very clear exposition of the Dirac equation including the quaternion rewrite of the Dirac equation. This is an excellent introduction to particle physics for students prior to university, during university and after university courses in physics.

The Quaternion Dirac Equation

This small book (only 40 pages) presents the quaternion form of the Dirac equation. The neutrino mass problem is solved and we gain an explanation of why neutrinos are left-chiral. Much of the material in this book is drawn from 'The Electron'; this material is presented concisely and inexpensively for students already familiar with QFT.

An Essay on the Nature of Space-time

This small and inexpensive volume presents a view of the nature of empty space without the detailed mathematics. The expanding universe and dark energy is discussed.

Elementary Calculus from an Advanced Standpoint

This book rewrite the calculus of the complex numbers in a way that covers all division algebras and makes all continuous complex functions differentiable and integrable. Non-commutative differentiation is covered. Gauge covariant differentiation is covered as is the covariant derivative of general relativity.

Even Mathematicians and Physicists make Mistakes

This book points out what seems to be several important errors of modern physics and modern mathematics. Errors like the misunderstanding of rotation, the failure to teach the higher dimensional complex numbers in most universities, and the mathematical inconsistency of the Dirac equation and some casual errors are discussed. These errors are set in their historical circumstances and there is discussion about why they happened and the consequences of their happening. There is also an interesting chapter on the nature of mathematical proof within our society, and several famous proofs are discussed (without the details).

Finite Groups – A Simple Introduction

This book introduces the reader to finite group theory. Many introductory books on finite groups bury the reader in geometrical examples or in other types of groups and lose the central nature of a finite group. This book sticks firmly with the permutation nature of finite groups and elucidates that nature by the extensive use of permutation matrices. Permutation matrices simplify the subject considerably. This book is probably unique in its use of permutation matrices and therefore unique in its simplicity.

The Left-handed Spinor

This book covers the left-handed parts of mathematics which we call the chiral algebras. These algebras have CP invariance, violation of parity, and many other aspects which makes them relevant to theoretical physics. It is quite a revelation to discover that mathematics is left-handed.

Non-commutative Differentiation and the Commutator

(The Search for the Fermion Content of the Universe)

This book develops the theory of non-commutative differentiation from the fundamentals of algebra. We see what an algebraic operation (addition, multiplication) really is, and we discover that the commutator is a third fundamental algebraic operation within some division algebras. This leads to the first part of the derivation of the fermion content of the universe.

248

Index

#

$\mathbb{R}^n$, 5, 176

A_3 distance functions, 108

1

16-dimensional algebras, 86
1-dimensional trigonometric function, 34

2

2-dimensional rotations, 114

3

3-dimensional addition identity, 34
3-dimensional algebras, 27
3-dimensional angles, 32
3-dimensional curl, 150
3-dimensional distance functions, 112
3-dimensional parity, 62
3-dimensional trigonometric functions, 30

4

4-dimensional rotation, 3, 36
4-dimensional rotation matrix, 43
4-dimensional version of the curl, 146
4-potential, 154

8

8-dimensional algebras, 85

A

additive inverse, 20
adjoint representation, 219
aitch-bar, 80
algebraic isomorphism, 69
algebraic relations necessary for rotation, 40
algebraic relationship between physical constants, 83
angle product, 47
angular momentum operators, 202
anti-matter, 144, 165
anti-quaternions, 89, 165, 171

B

B-differential, 227
B-field, 133

C

charge of the electron, 71, 84
classical mathematics, 1
Clifford algebra, 75
Clifford product, 75
colour charges, 86
commutation relations, 89, 92
commutation relations, electromagetic tensor, 170
commutative rotations, 206
commutator, 71, 88
commutator algebra, 218
complex numbers, 12
complex numbers, scaled form, 15
continuity equation, 162
cross product, 48, 50
curl, 54, 127, 135
curvature, space-time, 110

Index

D

E

F

G

H

I

K

L

Index

T

tensor, 169
three generations of particles, 196
trigonometric addition relations, 34
types of angle, 5

U

unitary transformation, 100

V

vector angles, 33

vector field over the complex plane, 47
vector in 2-dimensional space-time, 230

W

weak force, 226
wedge product, 75

Z

zero characteristic, 12

www.ingramcontent.com/pod-product-compliance
Lightning Source LLC
Chambersburg PA
CBHW072302200526
45168CB00014B/142